Research Methodology in the Built Environment

T0075906

Built environment students are not always familiar with the range of different research approaches they could be using for their projects. Whether you are undertaking a postgraduate doctoral programme or facing an undergraduate or masters dissertation, this book provides general advice, as well as thirteen detailed case studies, from sixteen universities, in seven countries, to help you get to grips with quantitative and qualitative methods, mixed methods of data collection, action research and more.

Vian Ahmed is a Professor at the University of Salford and was previously director of postgraduate research. Her teaching expertise is in construction management and IT at undergraduate and postgraduate levels. She has broad research interests, covering skills culture and people in construction, for example: e-learning in construction, construction IT for large organisations and SMEs, ontology development for construction education, construction management and organisation readiness.

Alex Opoku is currently the Director for the Research Centre for Sustainability and Resilient Infrastructure & Communities (SaRIC) at the London South Bank University. He is a Senior Lecturer in Quantity Surveying and the Course Director for the MSc Quantity Surveying programme at the school of Built Environment & Architecture.

Zeeshan Aziz is leader of the Construction Management MSc at the University of Salford and also supervises undergraduate, postgraduate and doctoral research projects. His own research interests include intelligent construction collaboration, intelligent integrated sustainable construction and disaster resilient infrastructure.

Research Methodology in the Built Environment

A selection of case studies

Edited by
Vian Ahmed, Alex Opoku
and Zeeshan Aziz

Routledge
Taylor & Francis Group

LONDON AND NEW YORK

First published 2016
by Routledge
2 Park Square, Milton Park, Abingdon, Oxon OX14 4RN

and by Routledge
711 Third Avenue, New York, NY 10017

Routledge is an imprint of the Taylor & Francis Group, an informa business

British Library Cataloguing-in-Publication Data
A catalogue record for this book is available from the British Library

Library of Congress Cataloging in Publication Data
Names: Ahmed, Vian, editor. | Opoku, Alex, editor. | Aziz, Zeeshan, editor.
Title: Research methodology in the built environment: a selection of case
studies/Edited by Vian Ahmed, Alex Opoku and Zeeshan Aziz.
Description: New York: Routledge, 2016. | Includes bibliographical
references and index.
Identifiers: LCCN 2015040445 (print) | LCCN 2015044926 (ebook) | ISBN
9781138849464 (hardback: alk. paper) | ISBN 9781138849471 (pbk.: alk.
paper) | ISBN 9781315725529 (eBook)
Subjects: LCSH: Architecture – Study and teaching. |
Architecture – Research – Case studies.
Classification: LCC NA2000 .R48 2016 (print) | LCC NA2000 (ebook) | DDC
720.72 – dc23
LC record available at http://lccn.loc.gov/2015040445

ISBN: 978-1-138-84946-4 (hbk)
ISBN: 978-1-138-84947-1 (pbk)
ISBN: 978-1-315-72552-9 (ebk)

Typeset in Bembo
by Florence Production Ltd, Stoodleigh, Devon, UK

Contents

Illustrations

Figures

Tables

Contributors

Editors

Vian Ahmed is a Professor in the Built Environment with more than 20 years of experience in academia and industry. She is currently the Associate Dean International and the Director of the Online Doctoral Programme at the School of the Built Environment, University of Salford. Prior to that she was the Director of Postgraduate Research for more than 8 years at the same organisation. Her teaching expertise is in construction management and IT at undergraduate and postgraduate levels. She has a broad research interest covering skills culture and people in construction; for example: e-learning in construction, construction IT for large organisations and SMEs, data management, construction management and organisation readiness. She is a Senior Fellow of the Higher Education Academy, with well over 100 journal and conference publications and more than twenty graduated doctoral students and a number of continuing students. She holds a BEng (Civil), MSc (Construction) and a PhD in computer-aided learning in construction.

Alex Opoku is currently the Director for the Research Centre for Sustainability and Resilient Infrastructure & Communities (SaRIC) at the London South Bank University. He is a Senior Lecturer in Quantity Surveying and the Course Director for the MSc Quantity Surveying programme at the School of the Built Environment and Architecture. He joined the London South Bank University in 2013 having worked as Postdoctoral Research Associate at the University of Cambridge, Centre for Sustainable Development, in the Department of Engineering. Alex holds a PhD in Construction and Project Management from the University of Salford. Alex graduated from Nottingham Trent University with an honours degree in Building Management in 2005 and received an MSc degree in Quantity Surveying Commercial Management from the Leeds Metropolitan University in 2008. Dr Opoku also has a well-established academic track record in teaching and research in the sustainable built environment, with a number of publications in refereed conference proceedings and journal papers to his name. He believes in making academic research relevant to practitioners and has been engaging with the construction industry through workshops, seminars and conferences.

Zeeshan Aziz is currently working as Senior Lecturer in Construction Management and Programme Director of the MSc Construction Management programme.

His research interests are in the area of advanced IT techniques within integrated construction informatics environments. He has led and participated in various research projects in the areas of process innovation and information technology support for collaboration in preparedness, response and recovery during disasters. He has been involved in teaching at both postgraduate and undergraduate levels and has provided commercial consulting services in the area of process innovation.

Authors

Ahmad Taki is experienced in building physics and design strategies to promote and encourage sustainability. He is a member of the EPSRC College for the peer review of research proposals. He has already worked on several research projects, giving him considerable experience in managing research contracts and disseminating research findings. Dr Taki is an invited author of the 'Ch 3 Heat Transfer' section of the CIBSE Design Guide and was coordinator for REF2014 UoA16. He has successfully supervised ten PhD students and is currently supervising nine others. Dr Taki is a Subject Head for architectural technology and Programme Leader for MSc Architecture and Sustainability and BSc Architectural Technology.

Aisha Abuelmaatti is a Lecturer at Buckinghamshire New University. Her research interests encompass the trichotomy of people–process–technology in the areas of operations, project and supply-chain management. She has been a Course Leader within a private higher-education institution and had extensive involvement in quality assurance. She completed her PhD in 2012 at the University of Salford, focusing on e-readiness in small and medium-sized architecture, engineering and construction enterprises. Dr Abuelmaatti graduated with a BSc in Multimedia and Internet Technology in 2006 from the University of Salford. She obtained an MSc in Information Systems in 2007 from the same university.

Bulent Algan Tezel (BEng, MSc, PhD) is a civil engineer and a Research Fellow at the University of Salford, with more than 6 years of research experience. He is also currently acting as the committee Vice Chair and Board Secretary of the Lean Construction–UK North West chapter. Dr Tezel's research interests include lean construction, construction production system design, construction planning and control, construction process improvement and IT in construction. Before joining the University of Salford, he was involved in the delivery of EPC-based construction projects as a construction planning and control manager at a multinational construction group.

Christopher Barker is currently pursuing an MSc in Building Information Modelling and Integrated Design at the University of Salford. He is an experienced industry practitioner with extensive experience in design and project management.

David Baldry is a Senior Lecturer in the School of the Built Environment of the University of Salford, UK. He is a Chartered Building Surveyor and Chartered

Construction Manager, and his research interests include facilities management and the management of public-sector real estate.

David Greenwood is Professor of Construction Project Management in the Faculty of Engineering and Environment, Northumbria University in the UK. He is a director of BIM Academy (Enterprises) Ltd – a joint-venture spin-out company – and a former director of the Sustainable Cities Research Institute. He has published more than 100 conference and academic journal papers and has authored and co-authored several textbooks. He has more than 20 years of experience in consulting, training and lecturing around the world for commercial and governmental organisations and is an active promoter of initiatives that promote better practice in the industry.

Fidelis A. Emuze is Associate Professor and Head of the Department of the Built Environment and Head of the Unit for Lean Construction and Sustainability at the Central University of Technology, Free State, South Africa, where he teaches research methodology and construction management subjects. Lean construction, health and safety, supply-chain management and sustainability constitute the main construction research interests of Dr Emuze, who is a member of the Association of Researchers in Construction Management and the Lean Construction Institute.

Hasif Rafidee Bin Hasbollah is a Senior Lecturer in Wellness and Hospitality in the Faculty of Entrepreneurship and Business at University Malaysia Kelantan. He holds a PhD from University of Salford, Manchester, United Kingdom, in Facilities Management. His main research interests are in healthcare and hospitality facilities, particularly using the qualitative method.

Kola Ijasan is currently a Property Studies Senior Lecturer at the University of the Witwatersrand, Johannesburg, where he teaches real-estate corporate finance and valuation. Prior to his academic career, he was interested in property management and valuation. He has also undertaken consultancy and training for a range of private- and public-sector organisations, especially for market analysis and valuation methodologies. He is passionate about various topics in the built and human environment, especially housing market regeneration and urban housing. He has experience in supervising undergraduates and postgraduates, as well as professional students.

Jamal Alabid is currently a PhD student in the School of Architecture at De Montfort University. He completed his BSc and MA in 2004 and 2010, respectively, in Urban Planning and Architectural Design. He has worked for a consultancy bureau for 3 years and tutored at the Higher Technical College of Leptis Magna. He is competent with a number of 3D modelling software and simulation programs, such as ArchiCAD, Revit Architecture, Sketchup, Ecotect, DesignBuilder, SPSS and NVivo. His PhD project is investigating sustainable housing design in hot climates, with reference to Ghadames, Libya.

Julius Akotia is a lecturer in the School of Civil Engineering and Construction at Kingston University, London. He holds an MSc and a PhD in Construction Project Management. He has taught construction management courses at both undergraduate and postgraduate levels in UK higher education. He has also practiced in the UK construction industry. He has published more than eight international conference and academic journal papers. He has also peer-reviewed a number of conference and academic journal papers. He has a strong interest in research and has participated in major research projects.

Karim Ménacère (PhD, MA, BA, TEFL Dip) has broad teaching and research experience extending over 35 years as a university lecturer. He is currently a Senior Lecturer at the Liverpool Business School, Liverpool John Moores University, where he supervises PhD and DBA students in business, management and applied languages and translation studies (Arabic, French and English). He teaches research methods at the DBA level. He has supervised more than fifty doctoral students to successful completion and acts regularly as external examiner at other universities.

Lloyd Scott is Academic Advisor and Partnership Co-ordinator in the School of Surveying and Construction Management at Dublin Institute of Technology (DIT) and Professor of Practice at Oklahoma University. Apart from his lecturing, research and academic administrative duties, Lloyd teaches research methods to postgraduate students of the built environment at DIT. Along with this, he has produced many peer-reviewed conference and journal papers. He serves on the Editorial Board of the *International Journal of Construction Education and Research* and is the co-author of *The Assessment Toolkit*, a resource book for DIT academics. His research interests include modern approaches to thermal performance in domestic construction and development of sustainable energy sources and their practical application.

Mike Hoxley is Professor of Building Surveying at Nottingham Trent University. After training in local and central government, he worked for 15 years in private practice as an equity partner. Mike has been an academic for more than 20 years, completing his PhD at the University of Salford in 1999. He is a former editor of the journal *Structural Survey* and author of the RIBA *Good Practice Guide to Building Condition Surveys*. Mike's principal role at NTU is as postgraduate tutor for PhD students in the School of Architecture, Design and the Built Environment.

Monty Sutrisna is Associate Professor in Construction Management and is the Head of the Construction Management Department in Curtin University in Western Australia. As an active researcher, his research expertise includes construction procurement, construction IT and construction project management. He was the Director of Postgraduate Research Training and Outreach in the Research Institute for the Built and Human Environment in the University of Salford, UK, between 2007 and 2009 and has chaired the RICS COBRA Doctoral Session since 2008. His expertise in research methodology includes, but is not limited to, the application of grounded theory methodology techniques in built environment research.

Oliver Jones is a Senior Lecturer in Digital Architecture and Design Communication in the Faculty of Engineering and Environment, Northumbria University in the UK. Before entering academia, he founded and ran an architectural visualisation company. His current work deals with the integration of new technologies into university curricula and creating virtual learning environments for teaching, training and commercial purposes. His research focuses specifically on comparing patterns of visual behaviour and user experience between real and virtual environments. He has worked closely with commercial organisations in the field, including Crytek Gmbh (Germany), Enodo SA (France) and Autodesk Ltd (USA and UK).

Pathmeswaran Raju is a Reader in Knowledge Engineering and Deputy Head of the Centre for Knowledge-Based Engineering at the Birmingham City University (BCU). His current research interests include knowledge engineering for design, engineering and manufacturing disciplines for the development of knowledge-based engineering systems. His current role involves developing a knowledge-based decision support system for the INTERREG IVB NWE Energetic Algae project and knowledge models for the Platform Independent Knowledge Model project as part of the Strategic Investment in Low-carbon Engine Technology 2 programme and EU Clean Sky initiative. He joined BCU in September 2011 from the University of Salford, where he worked for more than 6 years.

Peter Barrett is Professor of Property and Construction Management in the School of the Built Environment at Salford University. To date he has produced more than 170 single-volume publications, refereed papers and reports and has made more than 110 presentations in around sixteen countries. From 2007 to 2010, he was President of the International Council for Research and Innovation in Building and Construction (CIB). He has undertaken a wide range of research, typified by a management focus on real-world problems using a range of hard and soft research methods, including experience in utilising grounded theory methodology.

Salman Azhar is a J.W. Wilborn Endowed Associate Professor in the McWhorter School of Building Science at Auburn University, Auburn, Alabama, USA. He has more than 20 years of research, teaching and construction industry experience working in the USA, Hong Kong, Thailand and Pakistan. Dr Azhar has conducted research on decision-support systems, high-performance buildings, building information modelling and construction safety. He has published more than 100 papers in refereed journals and at conferences. Dr Azhar is the Associate Editor of the *International Journal of Construction Education and Research*, USA, and an Editorial Board Member for the ASCE *Journal of Management in Engineering*.

Wisnu Setiawan has been a member of the academic staff in the Department of Architecture, Muhammadiyah University of Surakarta, Indonesia, since 2000.

Additionally, he has been involved in architecture, urban design and urban planning projects in Indonesia. He recently finished his PhD at the School of the Built Environment, University of Salford, UK. The research gave him experience in using a grounded theory methodology to explore the dynamic between the built environment and conflict-prone environment. It has sharpened his interest in sustainable design and development, particularly from a social perspective, including environmental psychology, community development, urban heritage and resilient urban environment.

Acknowledgements

As editors, we have been part of many research journeys over the years, as students and as academics. We have crossed many hurdles during our undergraduate, master and doctoral studies, either to complete a short dissertation or a thesis. Like most researchers, we have also hit a few brick walls at times, feeling stuck, frustrated and wanting to get off the train to stop the research journey. As academics, we have also been part of other people's research journey, knowing for sure that, if these journeys are fed with the right fuel, they will arrive at their final destination.

We would, therefore, like to thank all our mentors within the academic community and the community of practice in construction who helped us and supported us during our research journey, and all our undergraduate, postgraduate and doctoral students who have inspired our thinking to produce this book, for much-needed, but at times hated, aspects of research methodology.

We would also like to thank Routledge for helping us realise our long-term ambition to produce this book and understanding the existing need for sharing examples of good practices in research methodology.

We would also like to thank our families for allowing us to take time out of their entitlement to put this book together. Some missed out on a good meal for the day, others missed out on play time, and others sat watching us sitting with our digital friend with a few superficial conversations taking place. We sincerely thank you for your patience and dedicate this book to you.

Vian Ahmed
Alex Opoku
Zeeshan Aziz

Introduction

Over the past few years, the number of postgraduate researchers has witnessed a massive increase. These researchers arrive from all over the world, striving to do well in their studies. Research methodology seems to be one of the greatest stumbling blocks that researchers and the less converted supervisors encounter, with a great deal of inconsistency in the level of adaption and understanding of research methodology. This is mainly owing to the fact that most of the existing research methodology books only focus on the theoretical concepts, without clearly demonstrating how these concepts apply to real-life scenarios.

This book aims to guide researchers in the built environment through the thought process of identifying a sound research methodology and identifying a suitable method of data collection. To achieve this aim, the reader will be guided through show-cased quality research that demonstrates:

- a simplified understanding of research methodology, enriched with existing examples of built environment research to demonstrate the understanding of research philosophies, research approaches and research methods, down to identifying the unit of study;
- the thought process through which a sound methodology is selected;
- a comprehensive range of cases studies that cover various approaches to data collection such as: qualitative methods of data collection, quantitative methods, mixed methods, the grounded theory approach, design science research and action research.

The book consists of sixteen chapters split into the following seven parts:

- Part I: The thought process of the research journey
- Part II: Quantitative research
- Part III: Qualitative research
- Part IV: Mixed methods research
- Part V: Action research
- Part VI: Grounded theory research
- Part VII: Design science research

This book will help the reader view different case scenarios that have been formed by researchers in the built environment to answer specific or evolving research questions in a simple and easy way that students at masters and PhD levels can relate to. The book will address fundamental issues that researchers must trigger (or find ways of triggering) in order to adapt a sound research path. The edited book will also be useful for academics, researchers and practitioners.

Part I

The thought process of the research journey

1 Getting ready for your research

Setting the scene

Vian Ahmed and Alex Opoku

This chapter sets the scene for the book and will elaborate on types of research and the thought process of the research journey, from start to finish. The chapter produces a breakdown of three main zones: the literature review zone (how to be equipped to tackle it), the research methodology zone (how to approach this stage) and the data-collection stage (how to prepare for it). The chapter demonstrates the importance of the introductory chapter of any thesis; the thought process for shaping the introduction to the research (including the abstract) and how the introductory chapter of the thesis could be designed to give an overview of the research and set the scene for the thesis; how to shape the aim and objectives; and how to derive the research problem, which may lead to formulating a research hypothesis or the research questions.

1.1 Introduction

In its simplest terms, research is defined by the *Concise Oxford Dictionary* as 'the systematic investigation into and study of materials and sources in order to establish facts and reach new conclusions'. In its broadest terms, research is about gathering facts through information and data in order to advance existing knowledge, which is an essential element of academia and practice. Although research may differ in the way it is conducted across disciplines and in the value it generates over time, it remains a core component for taking academia and practice forward, creating a short-term or long-term impact on society, economy, policy, the environment and quality of life. Therefore, to clearly portray the essence of the research (whether at undergraduate or postgraduate level), it is important that researchers follow a logical and simplistic pattern for developing their research.

Given that the focus of this book is the built environment discipline, research on the built environment has been defined by Roof and Oleru (2008) as, 'the human-made space in which people live, work, and recreate on a day-to-day basis'. Therefore, the nature of research in the built environment is multidisciplinary, covering human and organisation management, technology and environment, to promote innovative thinking and problem-solving issues surrounding the environments created by humans and for human activities, whether it is to do with infrastructure projects, urban spaces or buildings. Accordingly, there are different methodologies that are being adapted to conduct research into the built environment. Despite the differences in the methodologies that have been used, this chapter lays out for the reader the research journey and the thought process that researchers would be expected to follow in order to reach definitive milestones within their research and progress with their research journey. This chapter, therefore, intends to give simple guidance on the thought processes covered by the following milestones:

- understanding the motivation for the research;
- defining the research gap;
- shaping the aim and objectives;
- synthesising the literature and defining the underpinning theory;
- selecting a suitable methodology;
- defining the data-collection strategy;
- articulating the research findings and ending the journey.

1.2 The motivation to research the research gap

It is important that researchers reflect on their own motivation to do the research and what inspired their thinking, which often sets the initial direction of the research. There are often different scenarios that capture the motivation for the research, for example:

- Researchers may be aware of the research problem from being immersed in the world around them; hence, this motivates them to explore the problem further in order to propose valid solutions.
- Researchers are curious about a specific knowledge area or a new technology and, therefore, are eager to explore it further in order to make a new discovery, or develop a new product.
- Researchers know that a problem exists, and they wish to prove that it does.
- Researchers want to explore new grounds of research, but have no idea where to start.

Whatever the scenario, the motivation for any research revolves around a clearly defined knowledge gap and a well-defined research problem. Therefore, one of the factors that contribute to weakly substantiated academic research is when the research gap is vague or the research problem is not strongly founded on literature, even if the researcher is adamant that the problem exists. Therefore, a good justification to why the research is needed is the first milestone that needs to be realised.

To identify the research gap and clearly define the research problem, here are some tips:

- *Look at the big picture*: Researchers should look at the big picture, conducting a thorough literature review in order to: understand the current state of the art, how the field has advanced over time and what are the old and current developments; identify any initiatives that have evolved over time; be aware of similar work in the field; identify the key players; and address their contribution. It is also very important for research that is focused on a particular country, or region, to draw from existing practices and advancement internationally, before narrowing down the research to focus on a smaller geographical scale. Another example is to do with research that is intended to develop a digital platform, where the researcher needs to explore other platforms in order to show that the proposed platform is sufficiently unique when compared and contrasted in principle with other similar digital developments, giving a well-justified reason to why the proposed development is needed.

- *Be factual*: Researchers should not make any assumptions that are merely based on their current knowledge and experience, to be used as evidence of the research gap or research problem. This knowledge or experience may have motivated their study, but if it is not backed up with sufficient evidence from the literature, then it is not factual. For example, if the researcher works in a construction firm and is adamant that there is a large amount of a certain material that is being wasted and disposed of on construction sites, based on their daily observations, it would be wrong to generalise such a research problem that is confined to this situation, if such facts cannot be cited from literature or be informed by reliable resources.

- *Display convincing evidence*: Researchers should back up their arguments with citation references, contrasting different views from literature and forming critical arguments and standpoints. It is important that the citation references are up to date to reflect the advancements in the field. It would, therefore, be wrong for researchers who are doing research in 2015 and, for example, addressing building information modelling as a new phenomenon to refer to Smith (2007). Such a citation is considered to be out of date for such an advanced topic.

- *Consider the writing style*: Researchers must try to be assertive in their arguments. To give a big picture, it is important that this is done by starting with the generic issues in the field and gradually narrowing down the argument to the specific issues in the field. During this process, it is important that researchers take the readers with them on the journey by providing a smooth build-up of arguments, with a progressive flow of information through connected sections and paragraphs, using assertive arguments and connecting words such as 'however', 'although', 'therefore', 'while', etc.). It is, therefore, very important that researchers train themselves to be more reflective and critical of the information provided in order to be able to compare and contrast views and select the most suitable propositions that will take the research forward.

- *Get inside the head of the reader and other researchers*: Don't assume that the reader knows what you are talking about. Researchers often assume that the reader should know certain acronyms, or definitions, or even that they are an expert in the subject area. Unfortunately, this is not always the case. Research should be presented in a way that paves the path for new researchers to understand their work. For example, if a researcher has referred to 'the Socratic method of teaching' in their literature, with the assumption that everyone should know what this method means, this could create a gap or an element of frustration in the reader's mind, should they not be familiar with this method (although it is well known worldwide), and it may stop other researchers from relating to aligned concepts.

- *Citation references and use of resources*: Researchers should use a variety of resources and citation references (journals, conference proceedings, books, reports, etc.) in order to increase the reliability of, and level of confidence in, the current research in terms of depth and breadth, so that the research gap is clearly defined, with strong evidence.

- *Setting the boundaries of the research*: Researchers may find more than one gap in a particular knowledge area of research, and the gaps may be closely linked. It is, therefore, important that researchers make sure the research problem is crystallised, by clearly defining it and drawing clear boundaries. This is different to the 'limitations' of the research. Limitations are the issues that emerge out of the research and are out of the researcher's control. This may translate into having limited access to people within an organisation or organisations, or limited documents or data. These limitations will then open up opportunities for future research and considerations as to whether the findings of the research can be generalised in view of the displayed limitations.

- *Contribution to knowledge*: In simple terms, what did the research add (or intend to add) to the body of knowledge that did not exist before? Researchers need to reflect on their own value added to the body of knowledge, which did not exist before the research was conducted, and how this added value filled the research gap that was identified and contributed to resolving the research problem.

Figure 1.1 shows how to view research with the researcher's eye, starting from the generic concepts and gradually narrowing down the research into the specifics, identifying the research gap and setting the intentions of the research. Researchers should not try and provide research solutions that are looking for a problem.

1.3 Setting the research aim and objectives

There are numerous resources that explain how researchers should go about forming their aim and objectives, for example Mary *et al.* (2014) and Thomas and Hodges (2010). Although these resources provide different definitions of the research aim and objectives, most agree that the aim of the research must be sharp, clear and concise. It is about what the research intends to deliver after it has been accomplished. For example, if the research intends to 'develop a strategic approach for effective leadership in construction', then this is what would be expected to be delivered. Most dissertation assessors and PhD examiners tend to read the research aim at the start of the dissertation or the thesis and check the conclusions at the end of the dissertation or thesis to see if the aim of the research has been achieved. It is surprising that many researchers fail to bridge the gap between the aim and conclusions in order to bring the research to a meaningful end.

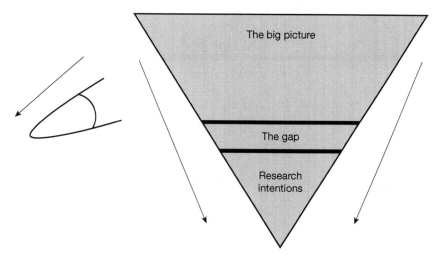

Figure 1.1 The research thought process through the researcher's eye

The objectives are directions that the researcher will take in order to achieve the aim of the research. The objectives need to be targeted towards measurable outcomes. They could lead to the development of an understanding of certain concepts, theories or definitions, for a given purpose – for example: 'To identify the underpinning theories of IT management and indicators of IT readiness in companies', 'To evaluate the factors that contribute to the implementation of collaborative technologies' or 'To draw up a set of recommendations that could be considered as a guide for organisations to implement IT'. Whatever the objectives are, they are intended to set a direction for the researcher to follow. Again, dissertation assessors and examiners tend to look for the objectives at the start of the dissertation or thesis and check these out at the end of these documents, to try and understand what directions the researchers followed to achieve the aim, and what tangible outcomes were delivered.

Although researchers are often advised to develop a research proposal before starting the research, and define the aim and objectives, these statements merely provide a starting point for discussion and constantly have to be revisited and refined, as the research progresses, and the research gap is clearly defined. Having a clear statement of aim and objectives will, therefore, inform the direction of the research towards achievable outcomes.

1.4 Research methodology

There are a number of resources that explain research methodology theories and principles and that are suitable for built environment research, with further explanation about research methodology in Chapters 2 and 3 of this book. The purpose of this section is to elaborate on the thought process that can help researchers to select a suitable methodology. It is, therefore, important that the researcher tries to find an answer to the following questions:

- Has the research gap been clearly identified? If so, what is it?
- Is the research problem clearly defined? If so, what is the nature of the research problem?
- How does the underpinning theory gathered from the literature inform the methodology, e.g.:
    - Have certain factors been identified that require testing to verify their existence?
    - Have certain questions been raised that need answering?
    - Has a specific problem been identified that needs to be explored?
    - Has the literature defined parameters that need evaluating?

It is, therefore, important that researchers build a bridge between the underpinning theory (what has already been established from the literature) and how that informs the methodology. The research methodology gives the researcher the opportunity to position their research problem in a suitable philosophy, develop a suitable approach to tackle the research problem, select a suitable research strategy that leads to appropriate methods for data collection, and tackle the right unit of study, ensuring the reliability and validity of the results.

1.5 Results, analysis, validity and reliability

Having defined a suitable methodology, approach, strategy and methods, it is important that researchers clearly state the following:

- What data-collection or data-generation tools are being used, and what are they being used for?
- How are the data being triangulated?
- What type of data is being collected, and why?
- From whom are the data being collected, and why have these sources been chosen?
- Where are the data being collected, and why?
- What is the sample size, and why is this perceived to be the right sample size?
- Are there any statistical tests being used, and, if so, why these tests?
- How will the data be analysed, and why?
- What observations are made upon the data gathered?
- What is the researcher's view of these observations?
- What are the main findings from these observations?
- What conclusions can be drawn from these findings?

Therefore, as researchers start to display the results of their data analysis, it is important that they are able to discuss the trends of the results and make observations on each data set in order to underpin the main findings. It is only through the accumulative list of findings that the researchers will be able to conclude their observations of the results.

1.6 Validity and reliability

In order to test the quality of the empirical data collected, researchers are expected to demonstrate the measures adopted to increase the data's validity and reliability. For example, there are a number of ways to go about validating the data, which should lead to some of the following questions:

- Has the researcher used multiple sources or a chain of evidence to validate the data, or perhaps the results were reviewed by a set of interviewees to confirm the validation?
- How did the research design and the data collected enable the research to draw accurate conclusions about the relationship between the cause and effect of the data obtained?
- What approach did the researcher adopt to generalise the findings of the research in relation to the domain of study?

Researchers should also be able to demonstrate the reliability of the results, that is, the extent to which, if the same study were repeated over a certain period of time or with different groups of people, it would give the same results. Researchers should also identify any bias or influential factors that may have impacted on, or distorted, the data and whether any measures have been taken to eliminate the likely bias.

1.7 Discussion, conclusions and recommendations

To complete the research journey, it is important that researchers are able to reflect on it and discuss the tangible outputs of their research. The outputs are usually directly related to the objectives of the research. This should enable researchers to pick the main findings from the outputs, whether they are from the gathered literature, the chosen methodology or the results obtained. Researchers should be able to observe the topology of these findings and form the conclusions of the research. Based on the derived conclusions, researchers should be able to make recommendations to the field to reflect the benefits of the study. There is no doubt that the added value of the research lies within the recommendations for future work, opening further gates for research building on existing findings.

1.8 Chapter summary

This chapter has attempted to give an overview of the thought process of the research journey in a very simplistic way, which can be summed up as elaborating on the big picture, identifying the research gap and clearly defining the goal of the research. There are other factors that should also be taken into consideration, such as the boundaries and limitations of the research, how to conduct the literature review, and what unique contribution to knowledge the research is making. One of the equally important aspects of the research involves clearly defining the underpinning theory, and how it informs the methodology, which then leads to the relevant choice of methods.

This book has been structured in a way that captures the research thought process through setting the scene, the research, consolidating the findings from the literature, and how these findings inform the methodology, which leads to different methods of data collection.

Bibliography

Bell, J. (2010). *Doing Your Research Project: A guide for first-time researchers in education, health and social science* (5th edn). Maidenhead, UK: Open University Press.

Concise Oxford Dictionary (1990). (8th edn). Oxford, UK: Clarendon Press.

Cryer, P. (2006). *The Research Student's Guide to Success* (3rd edn). Maidenhead, UK: Open University Press.

Fellows, R. F. and Liu, A. M. M. (2015). *Research Methods for Construction* (4th edn). Chichester, UK: John Wiley.

Fox, D., Gouthro, M. B., Morakabati, Y. and Brackstone, J. (2014). *Doing Events Research: From theory to practice* (1st edn). Abingdon, UK: Routledge.

Marshall, S. and Green, N. (2010). *Your PhD Companion: The insider guide to mastering the practical realities of getting your PhD* (3rd edn). Oxford, UK: How to Books.

Mary, D., Gouthro, B., Morakabati, Y. and Brackstone, J. (2014). *Doing Events Research: From theory to practice*. Abingdon, UK: Taylor & Francis.

Murray, R. (2009). *How to Survive Your Viva: Defending a thesis in an oral examination* (2nd edn). Maidenhead, UK: Open University Press.

Naoum, S. G. (2012). *Dissertation Research and Writing for Construction Students* (2nd edn). Oxford, UK: Elsevier.

Petre, M. and Rugg, G. (2010). *The Unwritten Rules of PhD Research* (2nd edn). Maidenhead, UK: Open University Press.

Philips, E. M. and Pugh, D. S. (2010). *How to Get a PhD: A handbook for students and their supervisors* (5th edn). Maidenhead, UK: Open University Press.

Roof, K. and Oleru, N. (2008). 'Public Health: Seattle and King County's push for the built environment', *Journal of Environmental Health*, 71: 24–7.

Saunders, M. N. K., Lewis, P. and Thornhill, A. (2012). *Research Methods for Business Students* (6th edn). Harlow, UK: Pearson Education.

Thomas, D. R. and Hodges, I. D. (2010). *Designing and Managing Your Research Project: Core skills for social and health research* (1st edn). London: Sage.

Yin, R. K. (2013). *Case Study Research: Design and methods* (5th edn). London: Sage

2 When more does not mean better

Selecting a research methodology and methods

Karim Ménacère

Novice researchers often experience a taxing and stressful time in their attempts to select the appropriate methodology and methods for their research. The extensive literature on the subject is overburdened by too much 'academism and obscurantism', compounded by conflicting positions, to the detriment of transparency and consistency in the use of user-friendly language. The tangle in the debate arises from a failure to recognise that some of the terms employed around key methodology and methods are used by authors in differing and often random and interchangeable ways. This chapter aims to highlight the academic inconsistencies and clarify the muddled jargon that generates key terminological and definitional issues. It seeks to guide novice researchers through the methodology maze, particularly as there is an increasing pluralism of alternative methodologies emerging, and to enable them to make a more informed decision about their methodological choices. It attempts to delineate the paradigm boundaries, seeking a connection between the ontology, epistemology, methodology and methods of each paradigm. In addition, it explores some underlying assumptions in research that impact on researchers' worldviews, theoretical frameworks and study designs. This chapter argues that defining methodology and paradigms is less important than understanding their nature and considering their functions. It aims at encouraging greater methodological awareness and understanding for better practice.

2.1 Introduction

Academic research is generally underpinned by basic philosophical assumptions, directly or indirectly determined and defined by the researcher. Yet, research students are often perplexed by the lack of clarity, the inconsistency and the wide variation of the language used to explain methodology and methods, paradigms, ontology and epistemology. Given the broad literature that already exists on research methods, one might be forgiven for thinking that there is little left to say, and yet that is precisely the point being made here. So many textbooks and journals on methodology and methods are written in a language that is either ambiguous or highly jargonistic, often using contradictory categorisations. There is a need to demystify and standardise the research terminology to make it accessible to a broader community of researchers. The research process focuses on the nature of the problem it aims to address and the type of information it seeks to find. Transparency and clarity in both the terminology and processes of methodology and methods are imperative. The current breadth and depth of material on research methodologies add to the confusion, rather than providing intelligibility. This makes the selection of an appropriate research design for a particular study a daunting task. Mkansi and Acheampong (2012: 132) refer to this predicament: 'Research philosophy classifications such as ontology, epistemology, and axiology and their conflicting applications to the "quantitative–qualitative" debates, are a major source of dilemma to research students'.

Often, the purpose of the methodology chapter is misunderstood by research students, who may, under pressure, compile a chapter of 20,000 words extracted from various popular research-method textbooks vaguely related to the objectives of the study under consideration. The resulting content of the chapter is unfocused and fragmented, not evidently linked to the research in question, and fails to explain and justify challenging key methodological concepts. Research students often treat, consciously or unconsciously, some of the current methodology textbooks like scriptures – truths that cannot be tampered with or questioned – and subsequently shoehorn material into their study that is alien to, or incompatible with, their research context and the problem they are addressing. They are heavily influenced by the prescriptive guidelines of the existing textbooks that overtly stipulate that research fits neatly into clear-cut categories such as a quantitative or qualitative approach, and each in turn is set firmly into its underpinning paradigm.

The overarching aim of this chapter is to untangle and make sense of research methodology and methods, providing an alternative view and highlighting the key issues and sources of confusion that pervade the current methodology debate. The intent is not to write another research methodology and methods manuscript: this topic has received and continues to receive considerable and skilled attention from the experts in this field (Guba and Lincoln, 1994; Mertens *et al.*, 2010; Bryman, 2012; Easterby-Smith *et al.*, 2012; Saunders *et al.*, 2012; Creswell, 2013; etc.). It simply intends to identify the nature of the problem new researchers are experiencing and the need for clear and consistent terminology regarding research methodology. This chapter consists of three parts: the first formulates the nature of the problem of the theoretical construct, the second explains the need for formulating clear constructs in methodological research, and the third explores how methodological labels mean different things to different researchers and vary across the epistemological and methodological spectrum.

2.2 The importance of methodology and the relevance of research

Although research is crucial to both business and academic enterprise, there is little consensus in the literature on how it should be defined: it means different things to different stakeholders. For the purpose of this chapter, research is a process of enquiry generating new ideas that make a difference, that have an impact and lead to change. Research starts with identifying the nature of a problem that is worth addressing and seeks to find a solution for it. Thus, knowledge generation is the essence of research. Outlining a research methodology helps others know what the research is trying to find out, why a particular piece of research is worth undertaking and why is it being conducted in a particular way. Doing research must, therefore, be methodologically based in order to generate all forms of knowledge and provide the means whereby understanding is created. A research methodology is often referred to as the theoretical underpinning of the research. It sets the directions of the research and the possible implications of the research. The methodology is also shaped by the literature review. To be fit for purpose, research findings must be founded on a clear methodological framework in order to be readily translatable into action. Research is not simply the adding of factual information to an existing body of facts. Jenkins and Zetter (2003: 11) assert that, 'Research is context specific and multidisciplinary rather than being hypothesis led; it uses fuzzy, rather than empirically based data; it is problem solving rather than deductive'. In short, research is the pursuit of knowledge gained through methodological processes of collecting and analysing information to enhance an understanding of a phenomenon. It aims to address a research question or questions in order to generate and develop knowledge.

Methodology and methods: clarifying differences and acknowledging similarities

Research methods and research methodology are two terms that are confused as one and the same and are often used interchangeably (Collis and Hussey, 2003). Strictly speaking, they are different, one of the main differences being that research methods are instruments for the collection of data about a phenomenon. Methods consist of the different investigation techniques, data-collection tools such as questionnaires, interviews, etc. In contrast, methodology is the study of methods and deals with the philosophical assumptions underlying the research process. Hussey and Hussey (1997: 54) define methodology as, 'the overall approach to the research process, from the theoretical underpinning to the collection and analysis of data'.

Methodology thus refers to the interrelationship that exists between theory, method, data and the phenomena under investigation. It is a roadmap that provides a clear vision and directives on how the research is to be conducted. Saunders *et al.* (2012) view methodology as a theory of (1) how research should be conducted and (2) the implications of the method(s) used by the researcher, given the assumptions on which the research is based. Kothari (2010) considers methodology as a systematic approach to solving a research problem. Crotty (2003: 3) believes that methodology is 'the strategy, plan of action, process or design', and method is 'the techniques or procedures used to gather and analyse data'.

Clarity in research methodology is paramount. Usage of terms and concepts whose meaning is indeterminate or ambiguous complicates the task of selecting appropriate methods and achieving useful research findings to benefit others. In essence, methodology is the theory that influences and shapes the researcher's ideas about which research methods will be used. It is the assumption that underpins the decisions made about the researcher's range of choices of what, who, where and how to investigate and the knowledge and information sought. Methodology, for Holden (2004: 6), 'is the researcher's tool-kit – it represents all the means available to social scientists to investigate phenomena'.

Student researchers' methodology chapters frequently contain misunderstandings of concepts, misuse of terminology and random application of methods. This results in uncertainty about what the researcher is aiming to find out, and the purpose of the research appears vague and unfocused. Moreover, there is often a discrepancy between method claims and the actual use of methods in students' research. This is caused essentially by blindly following the methodology and methods literature, thereby limiting the scope for innovation as the same research methods and designs are repeatedly and faithfully applied in ways that often restrict what is researched and how it is researched. Students select standardised, 'off-the-shelf' research instruments that fail to capture pertinent variables. Published materials on methodology and methods can exhibit lack of rigour, absence of clarity, and inconsistency in terms of definitions and classifications of fundamental concepts in research design. McGregor and Murnane (2010: 421) reinforce this view: 'the use of the terms paradigm, methodology and method is incredibly messy in the literature'. Crossan (2003: 64) claims that the 'distinction between *quantitative and qualitative philosophies* and research methods is sometimes overstated'. Crossan's wording illustrates the misuse of semantics, suggesting that *quantitative* and *qualitative* are philosophies rather than approaches or methods, as most researchers would agree. Similarly, Richards's book entitled *Qualitative Inquiry in TESOL* uses 'qualitative paradigm' (2003: 36). Thus, philosophical tags are arbitrarily attached to methodological terms without due explanation of the distinction between methodology, paradigm and methods. Broom and Willis (2007: 17) also seem to confuse the terms methodology and method as being one and the same. Spot the confusion in the following example:

> An interpretivist researcher will maintain that knowledge is socially constructed and reality is ultimately subjective; a positivist paradigm, however, will maintain that reality is fixed and that objective knowledge can be produced through rigorous methodology. *Methodologies utilised by the former are referred to as qualitative and those by the latter as quantitative* [italics added by the author].

Strictly speaking they are not, as qualitative and quantitative are methods not methodologies; methods usually refer to the techniques or instruments used to collect data, whereas methodology has a broader philosophical meaning that involves epistemological and ontological positions.

Gray (2014: 3, 5; italics added by the author) uses the terms 'inductive and deductive *reasoning*', 'inductive and deductive *methods*' and 'inductive and deductive *processes*', suggesting that methods, reasoning and processes are one and the same. Similarly, Sobh and Perry (2005: 1; italics added by the author) state that, 'Many researchers are concerned with the choice between a *quantitative and a qualitative methodology* . . . Some researchers use only one type of methodology while others suggest that both types may sometimes be appropriate'. Such indiscriminate use of terms has led many to voice serious methodological concerns. Bryman (2014) believes this research methodology 'mayhem' is creating a 'methodological ghettoization'. The same concern is reiterated by Mkansi and Acheampong (2012: 132):

> A number of studies have used different descriptions, categorisations and classifications of research paradigms and philosophies in relation to research methods with overlapping emphasis and meanings. This has not only resulted in tautological confusion of what is rooted where, and according to whom, but raises a critical question of whether these opposing views are enriching knowledge or subtly becoming toxic in the field.

Although often underestimated, the writing of the methodology and methods chapter is the most complex or obscure task, due to the care needed and the challenge of using meaningful but plain and consistent language, to demonstrate that the researcher possesses a firm grasp of the methodological concepts and that an informed choice is being made in selecting the appropriate underpinning paradigms. The researcher must justify the chosen research strategy in relation to the research objectives and explain the exclusion of alternative strategies. Discussion is needed of how ontological and epistemological choices are converted into specific methodological approaches, to reflect the researcher's methodological decision to opt for a particular method.

The conflation of terms in the textbooks – for example, methodology and methods, paradigm and methodology, qualitative and quantitative – seems ever more frequent, confusing assumed meanings with original meanings. This leads student researchers to insecurity and an inability to explain confidently, and justify beyond reasonable doubt, their methodological stance and the specific method-ological strategies derived from the ontological and epistemological assumptions that informed their decision. This lack of homogeneity of terminology around methodology and methods language results in researchers becoming hard-pressed to say what they mean and mean what they say. The accuracy and clarity of the terminology and language used are as important as the quality of an idea itself. There is a need for distinct concepts and definitions for clear and accurate thought, as the pertinence of research is based on organised and meaningful ideas. For many research students, the methodology and methods chapter is the most uncomfortable and frustrating phase of the doctoral journey. There is sometimes the mistaken assumption that the research methods section or chapter means simply going through the motions, reducing it to a process of rummaging through textbooks to gather as much information as possible, resulting in a 'copy and paste' exercise, with little thinking or understanding of the basic function of methodological concepts. It is not unknown for doctoral students to write the methodology chapter before conducting their literature review, because they are insufficiently trained or experienced to fully grasp the rationale behind such an important chapter. The methodological choice is decided, and the chapter is wrapped up, with little understanding or substantiation of the core philosophical underpinnings.

Another source of confusion is the recent proliferation and diversity of publications in research methodology and methods (e.g. Johnson and Onwuegbuzie, 2004; Biesta, 2010; Greene and Hall, 2010; Johnson and Gray, 2010; Mertens *et al.*, 2010; Denzin, 2012; Freshwater and Cahill, 2012). This adds to the lack of consensus among authors on fundamental paradigm issues and disagreements over methods. The boom in literature on research methods has often complicated rather than enlightened early researchers. In some of the literature, there is a lack of a consistent and coherent way of elucidating and labelling key methodological concepts, leaving the guesswork with research students, who then perpetuate the confusion between paradigm, philosophy and worldview. The key issue that has emerged from many recent methodology publications is inconsistency in methodological perspectives, which may be summarised as follows:

- a confusion between methodology and methods;
- a lack of distinction between the terms paradigm, philosophy, epistemology and ontology, terms that are often used interchangeably;
- a mix-up between research strategy and research design;
- the conflation of research design and methods; and
- underestimation of how methodological perspectives can play a crucial role in research.

More methodology textbooks should use unambiguous and consistent terms that novice research students find understandable and can trust to be accurate. The materials need to explain and articulate clearly key concepts and their functions in research.

2.3 Distinguishing research design from research method

Research design and research methods are often jumbled together, and yet research design is distinct from research method. A method is simply a means or technique of data collection, whereas a research design refers to the research arrangements and plans the researcher makes for how they intend to conduct research and collect and analyse data, in line with the purpose of the study and research objectives. Many research methods textbooks make no distinction between research design and methods. Thus, research design is often thought of as an instrument of data collection rather than as a logical research path of enquiry. The research design is the master plan that determines and decides the methods and techniques for collecting and analysing the required data. However, there is nothing inherent in any research design that requires or forces the researcher to adopt a particular method of data collection. Selecting the most fitting research design in any study is largely contingent on the objectives of the research and the nature of the problem the study aims to address.

The relevance of research design

Research design has a direct bearing on the reliability of the results the researcher aims to achieve. It provides a solid platform for the entire research, and, therefore, a clear research design is imperative, as it will facilitate the conducting of the research operation. Failure to discriminate between design and method leads to inadequate evaluation of designs. In short, a research design is a framework used for the planning of the research journey. It is a systematic structure of what is to be done, how it will be done and how the data will be analysed.

2.4 Rethinking philosophical foundations

Research philosophy, research paradigm and worldview are usually put under the same umbrella, suggesting these are just different labels signifying the same thing. Moreover, philosophy, paradigm, worldview, ontology and epistemology are presented as purely theoretical abstractions, of complex intellectual interest but detached from the real world. Student researchers become baffled by the way philosophical foundations such as philosophical assumptions, worldview and paradigm, which constitute a set of beliefs and assumptions about the nature of knowledge, are ill defined. Many authors make the assumption that their readers are aware of, and understand, the nuances of meaning of the various terms they are using. There is little evidence that indicates, explains or justifies from the outset why *paradigm* is used interchangeably with *philosophy* by many writers. There are diverse definitions of the term *paradigm*, but little is said about research philosophy or worldview and their link with each other. Thus, the multiple and broad perspectives of philosophy and paradigm are, for many, unintelligible, as these are used inconsistently.

Research philosophy

Understanding the philosophical perspectives underpinning research, and how these perspectives influence research methods and findings, is a crucial part of the research process. The research philosophy underpinning any study stems from key assumptions made about how reality is viewed. These assumptions help shape the research methods chosen as part of the research methodology. In essence, research philosophy refers to the nature and development of knowledge. The philosophy can also be defined in terms of the search for truth. So, where does that leave the term paradigm? Is it a clone of philosophy? Paradigm is a term that causes quite a stir and has attracted the attention and publication of some interesting and some controversial papers, occasionally reflected in title choices. For example, Hussain *et al.*'s paper (2013), 'Research Paradigms: A slippery slope for fresh researchers', argues that paradigms are not just a stumbling block for novice researchers but an unpredictable journey. Elshafie's article (2013), 'Research Paradigms: The novice researcher's nightmare', echoes the same thought, stressing that paradigms are a nightmare for research students, and Mkansi and Acheampong (2012) speak of paradigm as the 'students' dilemma'.

Worldview as a philosophical foundation

A worldview (or model of reality) can be regarded as a complex medium of beliefs and assumptions that enable individuals to make sense of and interact with the world around them. Guba and Lincoln (1994: 105) consider that worldview is the 'basic belief system or worldview that guides the investigator'. For Olsen *et al.* (1992), the concept of worldviews is that they are the mental lenses or embedded ways of perceiving the world. The term worldview is a mental image of reality, a set of general perceptions about the world and the position of an individual in the world. For Crotty (1998), worldview or philosophy refers to attitudes and beliefs about knowledge. Often, writers make little distinction between paradigm and worldview, considering a paradigm to be a basic set of beliefs and assumptions, whereas a worldview underpins the theories and methodology of a subject. Thus, like philosophy, worldview denotes a set of beliefs. However, the concept of 'worldview' is broader than the term 'philosophy', as the first includes the second. By the same token, a paradigm is more general than a philosophy, but narrower and more focused than a worldview. Worldview and philosophy represent in-depth levels of reflection of the world. According to Hart (2010: 2), 'worldviews are cognitive, perceptual, and affective maps that people continuously use to make sense of the social landscape and to find their ways to whatever goals they seek'.

In short, a worldview is a vision of reality and a mindset concerning what existence is all about. In plain and accessible narrative, the terms paradigm, philosophy and worldview denote a shared lens through which people look but see different things. Though related, the differences between these terms are a matter of degree. The conclusion that can be drawn from the above philosophical foundations is that they share common features in terms of their generic meaning. The definitions differ in form and tone, but are interrelated in content. Vagueness still persists as to whether these philosophical assumptions can be viewed as a minimum set of assumptions that are considered as true or merely as the lens through which a specific object or phenomenon can be investigated. Thus, philosophy, paradigm and worldview are terms that overlap in meaning and purpose. This overlap inevitably leads to some confusion, as the distinctions are not explicitly stated, making difficult the research students' task to determine what is paradigmatically appropriate and whether these can be mutually compatible in a given specific context.

2.5 Paradigms: the rite of passage to research

The student researcher is advised that a firm grasp of the research paradigms is a prerequisite for any research journey. A paradigm, as the basic belief system, has a direct bearing on the researcher's choice of epistemology, ontology and method-ology. Academics stress that the first phase of conducting research is to determine the relevant research paradigm as the basis on which the methods are decided. In other words, an awareness of the practical implications of a research paradigm reinforces the researcher's understanding of what it is they are investigating and allows them to justify the philosophical belief that has been adopted. As Brannen (2005:7) argues, 'the researcher's choice of methods is said to be chiefly driven by the philosophical assumptions – ontological and epistemological – which frame the research or the researcher's frame of reference'.

Thus, the underlying assumption is that a paradigm is mandatory and can be likened to a work permit to conduct research. The importance of a paradigm cannot be overlooked, as there is more than one reason why knowledge of research paradigms is useful to the researcher. Easterby-Smith *et al.* (2012: 17) state that, first, an understanding of paradigms helps in explaining the research design. This not only 'involves considering what kind of evidence is required and how it is to be gathered and interpreted, but also how this will provide good answers to the basic questions being investigated in the research'. Second, knowledge of paradigms allows the researcher to know which strategies will work and which will not. Accordingly, it helps guide researchers. Third, it can help researchers to classify and even produce strategies that may be outside their previous experience. It may also enable them to adjust their research strategies according to the limitations of diverse subjects or knowledge structures. Paul and Marfo (2001: 532) highlight two important aspects of paradigms:

> The first is that paradigms differ in their assumptions about what is real, the nature of the relationship between the one who knows and what is known, and how the knower goes about discovering or constructing knowledge. The second is that paradigms shape, constrain, and enable all aspects of educational inquiry.

However, in the light of the many fuzzy and conflicting views of the term 'paradigm', this methodological concept is far from straightforward, as Prasad (2005: 8) states:

> In the material world of actual research practice, the tidy abstraction of the paradigm as a hermetic domain of shared assumptions and world-views quickly begins to give way to the messy reality of contested ideas, multiple ongoing influences, and constant experimentation.

In deciding a research methodology, Guba (1981: 76) suggests that, 'it is proper to select a paradigm whose assumptions are best met by the phenomenon being investigated'. Others argue that the choice of paradigm is seen as secondary to deciding the type of theory to be produced, in terms of selecting a research method. There is no 'one best philosophical way', 'rather what we have are competing philosophical assumptions that lead us to engage with [social phenomena] . . . in particular ways' (Johnson and Duberley, 2000: 4).

Fretting about which paradigm is suitable to research, according to Guba and Lincoln (1994: 105), is of secondary importance to questions of research methods; they stress that, 'both qualitative and quantitative methods may be used appropriately with any research paradigm. Questions of method are secondary to questions of paradigm'. In support of the above stance, Howe (1988:1) argues that no incompatibility between quantitative and qualitative methods exists at either the level of practice or that of epistemology, and that there are, therefore, no good reasons for researchers to fear forging ahead with 'what works'. Thus, although the researcher's paradigm positioning suggests that the selected method(s) is not random but based on a clear philosophical assumption, in terms of best practice, researchers should consider what works and what kind of knowledge they seek to generate.

Paradigms – who needs them? A synopsis of the ongoing debate

The paradigm battleground is not new: it dates back to the 1980s, and the clash was triggered by claims for the supremacy of one paradigm over others, asserting that some paradigms are better and more suited to achieve the desired outcome of the research. This debate, known as the 'paradigm wars', which was followed by an 'incompatibility theory', claims that quantitative and qualitative are irreconcilable owing to their very different underlying philosophies. The argument, however, seems flawed and often borders on triviality, as each paradigm has its own merits and demerits. Peshkin (1993: 36) argues that, 'No research paradigm has a monopoly on quality. None can deliver promising outcomes with certainty'.

As a response to the battle between advocates of the single paradigm versus the mixed paradigms, many researchers stress that the ontology and epistemology conundrum is unlikely to generate a consensus in the paradigm debate. It is not a linear process, where one philosophy stops and the other starts. Paradigm divisions and categories are a state of mind. Everyone creates their own worldview, and each in their own way tries to make sense of the world around them. Connell and Nord (1996:1) conclude that any philosophical debate is moot because '[we do] not know how to discover a correct position on the existence of, let alone the nature of reality'. Hughes and Sharrock (1997:13) suggest that providing any guidelines to an appropriate philosophical stance is difficult:

> Since the nature of philosophy, and its relationship to other forms of knowledge, is itself a major matter of philosophical dispute, there is, of course, no real basis for us to advocate any one view on these matters as the unequivocally correct conception of the relationship between philosophy and social research.

The paradigm conflict produces blurred arguments, which stifle creativity and inhibit the new researcher, who is stuck with two options: they can either take the quantitative or the qualitative road, but not both at the same time. The contentious issue, which is obfuscating research innovativeness, is the sustained claim of a clear divergence between quantitative and qualitative methods.

Methodology is not an end in itself, and the process of selecting a research methodology should neither be ruled nor decided by a paradigm influence. Selecting a specific methodology should be based on its fitness to answer the research questions, a choice determined and decided by the researcher's epistemological and ontological assumptions. Research is investigating and understanding phenomena and real experiences about reality, rather than adopting an 'off-the-peg' method.

Research is, thus, about selecting appropriate methods rather than relying heavily on the philosophical underpinning. Research methods should not be an either/or blueprint, but rather about using a holistic approach in order to gain a deeper understanding of the topic under consideration. Contributing to knowledge, which is the quintessence of any research, should be upheld. Too much emphasis on the technicalities implies that the research will be 'technique-driven and lose track of the real purpose in generating information' (Seymour, 1989: 27).

It is self-evident that research students are not nurtured to challenge or question the assumptions of the established methods, but, as these methods are predominantly what most academic studies select, this fait accompli often leads them to a situation in which they have no room to manoeuvre or think about alternative methods. Moreover, ontological and epistemological commitments may also blind them to alternatives, because they suggest that researchers are constrained to see the world in a particular way, rather than focus on the specific context under consideration and the knowledge they seek to generate.

Paradigms: what is all the fuss about?

The paradigm debate has generated some warmongering language: 'paradigm wars', 'conflict', 'purists', 'separatists'. Rossman and Wilson (1985) categorise the antagonists into three factions arguing over whether the two approaches can be (or should be) mixed: that is, the 'purists', the 'situationalists' and the 'pragmatists'. The purists claim that the qualitative and quantitative approaches should not be merged or combined, because they each have a divergent underpinning philosophy; the situationalists contend that nothing is set in stone when it comes to methodological selection, because the choice of method is partially determined by the nature of research, and there is no reason why alternative method choices cannot be considered (Rossman and Wilson, 1985); and the pragmatists act as pacifists and view the two approaches as being capable of simultaneously making the most of their strengths, putting them both to good use (Hathaway, 1995). Thus, on one side are the purists, who strongly argue that mixing paradigms and methods is a serious flaw in judgement. The situationalists take up the middle ground, maintaining that certain methods can be mixed in specific situations. In direct opposition to the purists are the pragmatists, who argue that the whole debate rests on a false dichotomy between qualitative and quantitative research on the basis of their differing paradigms and advocate for the efficient use of both approaches. Creswell (2009), meanwhile, considers the 'paradigm wars' debate as bereft of substance.

The stand-off in the paradigm debate suggests that research should not be governed by an approach monopoly. It is only decided by what the researcher aims to find. No matter what paradigm is selected, the facts and figures, words and meaning are pertinent terms and form part of the analytical research process. The formulation and articulation of a clear paradigm terminology are hampered by two factors: first, there is a die-hard group campaigning for the exclusive use of one paradigm, one methodology and one method, limiting its scope and effectiveness. Second, there is a profusion of emerging definitions of key paradigms that are so fuzzy and misleading they have often become meaningless, defeating the purpose of benefitting researchers. The scale of lexical ambiguities and the woolliness of some of the paradigm definitions are illustrated in Table 2.1.

The challenges of understanding many of the above diverse definitions should not be underestimated. Masterman (1970: 61) points out that Kuhn, who originally coined the term paradigm, has himself provided twenty-two different definitions of the same term. Many of the definitions contain loose use of language and do not hold a single, universally understood position. The same thing is being repeated, by different people, each merely adding to the confusion, often leading to conceptual impasse. Some of the above definitions are recycled, providing muddled explanations of paradigm. Burrell and Morgan (1979: 23) use opaque jargon to define paradigms as 'very basic meta-theoretical assumptions, which underwrite the frame of reference, mode of theorising and modus operandi of the social theorists who operate within them'. For Bogdan and Biklen (1982: 65), a paradigm is 'a loose collection of logically held together assumptions, concepts, and propositions that orientates thinking and research'. Denzin and Lincoln (2008: 22) define paradigms as the researcher's 'net' that holds the ontological, epistemological and methodological beliefs, and refer to the 'taken for granted' aspects of a paradigm. Similarly, Gliner and Morgan (2000: 17) suggest that, 'Paradigm is a way of thinking about and conducting a research. It is not strictly a methodology, but more of a philosophy that guides how the research is to be conducted'.

In view of the scale of inconsistency, some have already voiced concern that the paradigm conflict could simply be replacing one set of problems with another. Lather (2006: 52) implicitly seems to suggest '*vive la différence*': 'Neither reconciliation nor paradigm war, this is about thinking difference differently, a re-appropriation of contradictory available scripts to create alternative practices of research as a site of being and becoming'.

At this point, it is worth considering the way forward in the paradigm debate, which has become passé and could be easily construed as a distraction. In the words of Buchanan and Bryman (2007: 486), there is a clear expression of let us move on: 'The paradigm wars of the 1980s have thus turned to paradigm soup, and organisational research today reflects the paradigm diversity of the social sciences in general'.

Admittedly, there is some common ground or consensus among authors – albeit each with their own jargon, which is often alien to people outside their exclusive circle – that a paradigm is a set of beliefs and assumptions about the world. Some authors have used evasive language to give the impression that they have something different or new to say about the way paradigms should be defined or understood. Dressing them up in new terms simply adds to research students' confusion, and many writers fail to realise that terms they know well may be difficult or meaningless to their audience: 'The difficulty in conducting research today is heightened by the incoherent classification of research philosophies such as epistemology, ontology,

Table 2.1 Authors' definitions of paradigm

Authors	Definition of paradigm
Kuhn (1962: 33)	'An integrated cluster of substantive concepts, variables and problems attached with corresponding methodological approaches and tools'
Kuhn (1970: viii)	'Universally recognised scientific achievements that for a time provide model problems and solutions to a community of practitioners'
Guba and Lincoln (1994: 105)	'A basic system or worldview that guides the investigator not only in the choices of method but in ontological and epistemological fundamental ways'
Guba and Lincoln (2005: 22)	'A net that contains the researcher's epistemological, ontological and methodological premises'
Chalmers (1982: 90)	'Made up of the general theoretical assumptions and laws, and techniques for their application that the members of a particular scientific community adopt'
Willis (2007: 8)	'A paradigm is thus a comprehensive belief system, worldview, or framework that guides research and practice in a field'
Mertens (2005: 7)	'A paradigm is a way of looking at the world. It is composed of certain philosophical assumptions that guide and direct thinking and action'
Neuman (2006: 81)	'A general organising framework for theory and research that includes basic assumptions, key issues, models of quality research, and methods for seeking answers'
Denzin and Lincoln (2008: 22)	'The net that contains the researcher's epistemological, ontological, and methodological premises may be termed a paradigm . . . All research is interpretive; it is guided by the researcher's set of beliefs and feelings about the world and how it should be understood and studied'
Harrits (2011: 152)	Paradigms 'refer to a common research practice, existing within a research community, and carrying with it a shared identity as a set of problems'
Blaikie (2010: 9)	'Assumptions made about the nature of social reality and the way in which we can come to know this reality'
Taylor *et al.* (2007: 5)	'A broad view or perspective of something'
Weaver and Olson (2006: 460)	'Paradigms are patterns of beliefs and practices that regulate inquiry within a discipline by providing lenses, frames and processes through which investigation is accomplished'
Denscombe (2008: 276)	A 'specific problem or set of problems that are regarded as particularly significant in relation to the advancement of knowledge'
Gliner and Morgan (2000: 17)	'Paradigm is a way of thinking about and conducting a research. It is not strictly a methodology, but more of a philosophy that guides how the research is to be conducted'
Bryman (1988: 4)	'A cluster of beliefs and dictates which for scientists in a particular discipline influence what should be studied, how research should be done, how results should be interpreted, and so on'
Maykut and Morehouse (1994: 4)	'A paradigm has come to mean a set of overarching and interconnected assumptions about the nature of reality. The word "assumptions" is key'

continued . . .

Table 2.1 Continued

Authors	Definition of paradigm
Bassey (1998: 8)	'A network of coherent ideas about the nature of the world and of the functions of researchers which, adhered to by a group of researchers, conditions the patterns of their thinking and underpins their research actions'
Collis and Hussey (2003: 46)	'Refers to the progress of scientific practice based on people's philosophies and assumptions about the world and the nature of knowledge'
Wilson (2001: 175)	'A set of beliefs about the world and about gaining knowledge that goes together to guide people's actions as to how they are going to go about doing their research'
Burrell and Morgan (1979: 23)	'Very basic meta-theoretical assumptions, which underwrite the frame of reference, mode of theorising and modus operandi of the social theorists who operate within them'
Bogdan and Biklen (1982: 65)	'A loose collection of logically held together assumptions, concepts, and propositions that orientates thinking and research'
Krauss (2005: 64)	'The identification of the underlying basis that is used to construct a scientific investigation'
Mills *et al.* (2006: 2)	'To ensure a strong research design, researchers must choose a research paradigm that is congruent with their beliefs about the nature of reality'
Fossey *et al.* (2002: 718)	'A system of ideas, or worldview, used by a community of researchers to generate knowledge'
Teddlie and Tashakkori (2009: 84)	'A worldview, together with the various philosophical assumptions associated with that point of view'
Creswell and Plano Clark (2007: 21)	A 'worldview'
Greene (2007)	Uses the term 'mental model' in much the same way as a worldview
Smith (1990: 43)	'The entire constellation of beliefs, values, and so on shared by members of a shared community'

Note: Compiled by the author

axiology and doxology and the quantitative–qualitative dichotomy debates' (Mkansi and Acheampong 2012: 132).

In conclusion, a paradigm refers to the different beliefs and ways of viewing and interacting with reality, and to the fact that research is affected and guided by a certain paradigm. Paradigms can be thought of as elements of one's imagination, serving as the lens through which the world is explored and interpreted.

Justification of research methodology and paradigm

One of the weakest links and commonest flaws in many studies is the lack of a clear methodological framework. As Caelli *et al.* (2003: 5) state, 'In the absence of an explicit methodology, the reader of these studies is left to speculate about the research approach, by piecing together clues based on data collection or analysis methods'.

Unsatisfactory methodology chapters result from regurgitating textbook material with little relevance to the study, 'forcing' fit rather than allowing fit to emerge (Duchscher and Morgan, 2004). Research students are told to discuss and justify why particular methodology and methods have been chosen to elicit data, but one of the snags with the justification approach is that the discussion may focus on endorsing one method and disparaging the others, in the sense of promoting one as being 'better'. Yet the selection of a method concerns fitness for purpose, and this is largely contingent on the research aims, and the choice is always a compromise between a number of factors, including validity, reliability and access to data and resources. It should be accepted that each method has its particular strengths and weaknesses.

Sandelowski (1993) warns researchers that being over-reliant on, or blindly defending, the methodology may result in overlooking methods that, after all, are responsible for creating knowledge and achieving key findings. As Johnson and Duberley (2000: 8) indicate, philosophical assumptions were 'usually left tacit or implicit and were rarely clearly presented or subjected to sustained reflection'. The methodology of a study is generally selected in line with the philosophical assumptions about reality and is shaped partly by the relevant literature, the nature of the problem and the research aim and question(s). Being over concerned with the selection and defence of methods, to the detriment of the essence of the research, has been referred to as 'methodolatry' (Chamberlain, 2000). Caelli *et al.* (2003: 5) stress that, 'We do not advocate in favour of idealising methodology but, rather, that relevant methodological issues and methods must be understood and clearly articulated'. Patton (1990: 39) advocates a 'paradigm of choices' that seeks 'methodological appropriateness as the primary criterion for judging methodological quality'.

This study argues that there is neither a rule nor a process to determine precisely which method fits better than another for a particular research. Research methodology is not designed in a vacuum or out of the blue: it is related and relevant to the study under consideration and dependent on the nature of the problem and the research question(s). Saunders *et al.* (2007: 116) believe that there is no one research philosophy better than the others: each is better at doing different things and, therefore, a researcher should select the right one that can help to achieve their research objectives. Which one is 'better' depends on the research questions the researcher is trying to answer:

> It would be easy to fall into the trap of thinking that one research approach is 'better' than another. They are 'better' at doing different things. Of course, the practical reality is that research rarely falls into only one philosophical domain.

Knowing the strengths and weaknesses of paradigms provides the researchers with insightful knowledge with which to decide their research positions. Drawing a clear demarcation between the two paradigms – positivism and interpretivism – is useful to understand the main differences between these two philosophies. Focusing wholly on one research philosophy, approach and strategy might reduce the effectiveness of the study, as Ozanne and Hudson (1989: 1) stress: 'blind conversion to interpretivism is just as dangerous as blind adherence to positivism'. In addition, considering quantitative and qualitative research as polar opposites is misleading, as Vogt (2008: 56) explains, and the 'quant–qual distinction distracts us from consideration of more important issues, and tends to constrain opportunities for innovation'. Understanding the social reality, researchers basically tend to mix quantitative and qualitative research techniques, methods, approaches, concepts or language in a single study. Gilbert (2008: 139) argues that, 'research that explicitly mixes paradigms leads to a fuller understanding of the social world . . . because complexity itself consists of both "interpretivist" and "positivist" aspects'.

According to Patton (1990: 30):

> Rather than believing that one must choose to align with one paradigm or the other, I advocate a paradigm of choices . . . the issue then becomes not whether one has uniformly adhered to prescribed canons of either logical-positivism or phenomenology but whether one has made sensible methods decisions given the purpose of the inquiry, the questions being investigated.

The rationale for undertaking research is to produce a story that stands up to close scrutiny and presents convincing and reliable evidence that can make a difference.

Tailoring the method to the nature of the study

If research methods are selected to address a research problem or a question, the debate over which method is best is avoided entirely. Fundamentally, qualitative and quantitative methods may seem alien to each other, are said to be different and are not easily combined. However, used intelligently, the two can compensate for each other's weaknesses and complement each other's strengths. According to Slife *et al.* (1999: 29), the rationale for selecting particular research methods is neither rule-driven nor objective:

> Method is not a transparent window or an objective instrument for testing our ideas. All methods (and languages) come with their own liabilities and assets and their own assumptions and implications. As a consequence, each method must be evaluated in relation to the context of its proposed use.

Paradigms or worldviews cannot be compartmentalised into frozen patterns of reality. In other words, reality cannot be rigidly polarised or categorised into different layers. Hamel *et al.* (1993: 28) state that the methodological issues 'cannot be appropriately considered, or for that matter resolved, by the opposition of quantitative and qualitative methods'.

2.6 Conclusion

To date, there are copious numbers of books and papers that have been published on research methods. This proliferation and increasing availability of research methods material has definitely provided useful insights and allowed research to be conducted on many levels and with degrees of complexity, but it has also generated some fundamental challenges for those who are involved in this process, including academics and research students. Some difficulties stem from the lack of simple and consistent use of language, which induces the mental fog hampering researchers in their efforts to select with confidence the research methodology and methods, particularly relating to emerging, alternative methods of conducting research. Too many inadvertently contain sweeping statements or make assumptions about key concepts.

Looking at it positively, the diversity and wealth of views that emerge from the methodology and methods literature may be perceived as signs of research maturity and tolerance, and that everyone is free to think that methodology and method are one and the same as long as the job gets done. However, in an attempt to bridge this diversity, confusion and ambiguity, it is proposed that the terms philosophy, paradigm, epistemology and ontology, rather than being used interchangeably, should be used with greater care and attention. Efforts should be made to use plain language that a lay person could understand, because grasping the meaning of these concepts will help student researchers write a meaningful research methods chapter, well suited to their particular study and crucial to producing a worthwhile piece of research.

The ontological, epistemological and methodological underpinnings of any study are concepts that are open to debate: they are not immutable. What social scientists study, what can be known about the topic of enquiry and how the research process should be conducted in order to create or generate knowledge have always been sources of contention. Research methods terms are often used synonymously, randomly or interchangeably to explain, imprecisely in some cases, philosophical foundations, including philosophical assumptions, worldview and paradigm. The justification for selecting a paradigm is 'fitness for purpose' – different research paradigms are suitable for different research questions and objectives. The ambiguous nature, inconsistent use and diverse shades of meanings of methodological perspectives and paradigms are seen as the basic flaws in research methodology and methods. This is not denigrating or suggesting that the various methodology and methods nomenclatures are superfluous. It simply views the inconsistency of some of the terminology used in textbooks as adding to the confusion and doing a disservice to the important subject of methodology and methods.

References

Bassey, M. (1998). Fuzzy generalisation: An approach to building educational theory. A paper presented at the British Educational Research Association Annual Conference, The Queen's University of Belfast, 27 August 1998.

Biesta, G. (2010). Pragmatism and the philosophical foundations of mixed methods research. In A. Tashakkori and C. Teddlie (eds), *Handbook of Mixed Methods in Social and Behavioural Research* (2nd edn, pp. 95–117). Thousand Oaks, CA: Sage.

Blaikie, N. (2010). *Designing Social Research*. Cambridge, UK: Polity Press.

Bogdan, R. C. and Biklen, S. K. (1982). *Qualitative Research for Education: An introduction to theory and methods*. Boston, MA: Allyn & Bacon.

Brannen, J. (2005). 'Mixing methods: The entry of qualitative and quantitative approaches into the research process', *The International Journal of Social Research Methodology*, 8(3): 173–85.

Broom, A. and Willis, E. (2007). Competing paradigms and health research. In M. Saks and J. Allsop (eds), *Researching Health: Qualitative, quantitative and mixed methods* (pp. 16–30). London: Sage.

Bryman, A. (1988). *Quantity and Quality in Social Research*. London: Unwin Hyman.

Bryman, A. (2012). *Social Research Methods* (4th edn). Oxford, UK: Oxford University Press.

Bryman, A. (2014). Quality Issues in Mixed Methods Research: Should We Bother with Research Methods Textbooks? Lecture delivered at University of Leicester, 24 February 2014.

Buchanan, D. and Bryman, A. (2007). 'Contextualizing methods choice in organizational research', *Organizational Research Methods*, 10(3): 483–501.

Burrell, G. and Morgan, G. (1979). *Sociological Paradigms and Organizational Analysis* (pp. 1–37). London: Heinemann.

Caelli, K., Ray, L. and Mill, J. (2003). '"Clear as mud": Toward greater clarity in generic qualitative research', *International Journal of Qualitative Methods*, 2(2): 1–24.

Chalmers, A. (1982). *What is this Thing Called Science?* Brisbane, QLD: Queensland University Press.

Chamberlain, K. (2000). 'Methodolatry and qualitative health research', *Journal of Health Psychology*, 5(3): 285–96.

Collis, J. and Hussey, R. (2003). *Business Research: A practical guide for undergraduate and postgraduate students*. Basingstoke, UK: Palgrave Macmillan.

Connell, A. F. and Nord, W. R. (1996). 'The bloodless coup: The infiltration of organization science by uncertainty and values', *Journal of Applied Behavioural Science*, 32(4): 407–27.

Creswell, J. W. (2009). *Research Design: Qualitative, quantitative, and mixed methods approaches* (3rd edn). Thousand Oaks, CA: Sage.

Creswell, J. W. (2013). *Qualitative Inquiry and Research Design: Choosing among five approaches* (4th edn). London: Sage.

Creswell, J. W. and Plano Clark, V. L. (2007). *Designing and Conducting Mixed Methods Research*. Thousand Oaks, CA: Sage.

Crossan, F. (2003). 'Research philosophy: Towards an understanding', *Nurse Researcher*, 11(1): 46–55.

Crotty, M. J. (1998). *The Foundations of Social Research: Meaning and perspective in research process*. Crows Nest. NSW: Allen & Unwin.

Crotty, M. J. (2003). *The Foundations of Social Research Meaning and Perspective in the Research Process*. Crows Nest, NSW: Allen & Unwin.

Denscombe, M. (2008). 'A research paradigm for the mixed methods approach', *Journal of Mixed Methods Research*, 2: 270–83.

Denzin, N. K. (2012). 'Triangulation', *Journal of Mixed Methods Research*, 6(2): 80–8.

Denzin, N. K. and Lincoln, Y. S. (2008). *Strategies of Qualitative Inquiry*. Thousand Oaks, CA: Sage.

Duchscher, J. B. and Morgan, D. (2004). 'Grounded theory: Reflections on the emergence vs. forcing debate issue', *Journal of Advanced Nursing*, 48(6): 605–12.

Easterby-Smith, M., Thorpe, R. and Jackson, P. (2012). *Management Research*. London: Sage.

Elshafie, M. (2013). 'Research paradigms: The novice researcher's nightmare', *Arab World English Journal*, 4(2): 4–13.

Fossey E., Harvey C., McDermott, F. and Davidson, L. (2002). 'Understanding and evaluating qualitative research', *Australian & New Zealand Journal of Psychiatry*, 36: 717–32.

Freshwater, D. and Cahill, J. (2012). 'Why write?' *Journal of Mixed Methods Research*, 6(3): 151–3.

Gilbert, N. (2008). *Researching Social Life* (3rd edn). London: Sage.

Gliner, J. A. and Morgan, G. A. (2000). *Research Methods in Applied Settings: An integrated approach to design and analysis*. Mahwah, NJ: Lawrence Erlbaum.

Gray, D. E. (2014). *Doing Research in the Real World* (3rd edn). London: Sage.

Greene, J. (2007). *Mixed Methods in Social Inquiry*. San Francisco, CA: Jossey-Bass.

Greene, J. and Hall, J. (2010). Dialectics and pragmatism: Being of consequence. In A. Tashakkori and C. Teddlie (eds), *Handbook of Mixed Methods in Social and Behavioural Research* (pp. 119–43). Thousand Oaks, CA: Sage.

Guba, E. G. (1981). 'Criteria for assessing the trustworthiness of naturalistic inquiries', *Educational Communication & Technology Journal*, 29: 75–91.

Guba, E. G. and Lincoln, Y. S. (1994). Competing paradigms in qualitative research. In N. K. Denzin and Y. S. Lincoln (eds), *Handbook of Qualitative Research* (pp. 105–17). Thousand Oaks, CA: Sage.

Guba, E. G. and Lincoln, Y. S. (2005). Paradigmatic controversies, contradictions, and emerging confluences. In N. K. Denzin and Y. S. Lincoln (eds), *Handbook of Qualitative Research* (pp. 191–215). Thousand Oaks, CA: Sage.

Hamel, J., Dufour, S. and Fortin, D. (1993). *Case Study Methods*. Newbury Park, CA: Sage.

Harrits, G. S. (2011). 'More than method? A discussion of paradigm differences within mixed methods', *Journal of Mixed Methods Research*, 5(2): 150–66.

Hart, M. A. (2010). 'Indigenous worldviews, knowledge, and research: The development of an indigenous research paradigm', *Journal of Indigenous Voices in Social Work*, 1(1): 1–16.

Hathaway, R. S. (1995). 'Assumptions underlying quantitative and qualitative research: Implications for institutional research', *Research in Higher Education*, 36(5): 535–62.

Holden, M. T. (2004). 'Choosing the appropriate methodology: Understanding research philosophy', *The Marketing Review*, 4(4): 397–409.

Howe, K. R. (1988). 'Against the quantitative–qualitative incompatibility thesis or dogmas die hard', *Educational Researcher*, 17(8): 10–16.

Hughes, J. A. and Sharrock W. W. (1997). *The Philosophy of Social Research*. Abingdon, UK: Routledge.

Hussain, M. A., Elyas, T. and Nasseef, O. A. (2013). 'Research paradigms: A slippery slope for fresh researchers', *Life Science Journal*, 10(4): 2374–81.

Hussey, J. and Hussey, R. (1997). *Business Research: A practical guide for undergraduate and postgraduate students*. London: Macmillan.

Jenkins, A. and Zetter, R. (2003). Linking Teaching and Research in Departments. Generic Centre/Learning & Teaching Support Network.

Johnson, P. and Duberley, J. (2000). *Understanding Management Research: An introduction to epistemology*. London: Sage.

Johnson, R. B. and Onwuegbuzie, A. (2004). 'Mixed methods research: A research paradigm whose time has come', *Educational Research*, 33(7): 14–26.

Johnson, R. B. and Gray, R. (2010). A history of philosophical and theoretical issues for mixed methods research. In A. Tashakkori and C. Teddlie (eds), *Handbook of Mixed Methods in Social and Behavioural Research* (2nd edn, pp. 69–94). Thousand Oaks, CA: Sage.

Kothari, C. R. (2010). *Research Methodology: Methods and techniques*. New Delhi: New Age International.

Krauss, S. T. (2005). 'Research paradigms and meaning making: A primer', *The Qualitative Report*, 10 (4): 758–70.

Kuhn, T. S. (1962). *The Structure of Scientific Revolutions* (2nd edn). Chicago, IL: University of Chicago Press.

Kuhn, T. S. (1970). *The Structure of Scientific Revolutions* (3rd edn). Chicago, IL: University of Chicago Press.

Lather, P. A. (2006). 'Paradigm proliferation as a good thing to think with: Teaching research in education as a wild profusion', *International Journal of Qualitative Studies in Education*, 19(1): 35–57.

McGregor, S. L. T. and Murnane, J. A. (2010). 'Paradigm, methodology and method: Intellectual integrity in consumer scholarship', *International Journal of Consumer Studies*, 34(4): 419–27.

Masterman, M. (1970). The nature of a paradigm in criticism and the growth of knowledge. In I. Lakatos and A. Musgrave (eds), *Criticism and the Growth of Knowledge* (pp. 59–90). Cambridge, UK: Cambridge University Press.

Maykut, P. and Morehouse, R. (1994). *Beginning Qualitative Research: A philosophical and practical guide*. London: Falmer Press.

Mertens, D. (2005). *Research and Evaluation in Education and Psychology: Integrating diversity with quantitative, qualitative, and mixed methods*. Thousand Oaks, CA: Sage.

Mertens, D., Bledsoe, K. L., Sullivan, M. and Wilson, A. (2010). Utilization of mixed methods for transformative purposes. In A. Tashakkori and C. Teddlie (eds), *Handbook of Mixed Methods in Social and Behavioral Research* (2nd edn, pp. 193–214). Thousand Oaks, CA: Sage.

Mills, J., Bonner, A. and Francis, K. (2006). 'The development of constructivist grounded theory', *International Journal of Qualitative Methods*, 5(1): 25–35.

Mkansi, M. and Acheampong, E. A. (2012). 'Research philosophy debates and classifications: Students' dilemma', *The Electronic Journal of Business Research Methods*, 10(2): 132–40.

Neuman, W. (2006). *Social Research Methods*. Boston, MA: Pearson.

Olsen, M. E., Lodwick, D. G. and Dunlap, R. E. (1992). *Viewing the World Ecologically*. San Francisco, CA: Westview Press.

Ozanne, J. L. and Hudson, L. A. (1989). 'Exploring diversity in consumer research', *Special Volumes – Association for Consumer Research*, pp. 1–9.

Patton, M. Q. (1990). *Qualitative Evaluation and Research Methods*. Newbury Park, CA: Sage.

Paul, J. L. and Marfo, K. (2001). 'Preparation of educational researchers in philosophical foundations of inquiry', *Review of Educational Research*, 71(4): 525–47.

Peshkin, A. (1993). 'The goodness of qualitative research', *Educational Researcher*, 22(2): 23–9.

Prasad, P. (2005). *Crafting Qualitative Research: Working in the postpositivist tradition*. Abingdon, UK: Routledge.

Richards, K. (2003). *Qualitative Inquiry in TESOL*. Basingstoke, UK: Palgrave Macmillan.

Rossman, G. B. and Wilson, B. L. (1985). 'Numbers and words: Combining quantitative and qualitative methods in a single large-scale evaluation study', *Evaluation Review*, 9: 627–43.

Sandelowski, M. (1993). 'Rigor or rigor mortis: The problem of rigor in qualitative research revisited', *Advances in Nursing Science*, 16(2): 1–8.

Saunders, M., Lewis, P. and Thornhill, A. (2007). *Research Methods for Business Students* (4th edn). Harlow, UK: Pearson Education.

Saunders, M., Lewis, P. and Thornhill, A. (2012). *Research Methods for Business Students* (7th edn). Harlow, UK: Pearson Education.

Seymour, D. (1989). 'Soft data—hard data: The painful art of fence-sitting', *Journal of Consumer Marketing*, 6(2): 25.

Slife, B. D., Hope, C. and Nebeker, R. S. (1999). 'Examining the relationship between religious spirituality and psychological science', *Journal of Humanistic Psychology*, 39(2): 51–85.

Smith, J. K. (1990). Alternative research paradigms and the problem of criteria. In E. Guba (ed.), *The Paradigm Dialog* (pp. 167–87). Newbury Park, CA: Sage.

Sobh, R. and Perry, C. (2005). 'Research design and data analysis in realism research', *European Journal of Marketing*, 40(11/12): 1194–209.

Taylor, B., Kermode, S. and Roberts, K. (2007). *Research in Nursing and Health Care: Evidence for practice*. Sydney, NSW: Thomson.

Teddlie, C. and Tashakkori, A. (2009). *Foundations of Mixed Methods Research*. Thousand Oaks, CA: Sage.

Vogt, W. P. (2008). 'Quantitative versus qualitative is a distraction: Variations on a theme by Brewer and Hunter (2006)', *Methodological Innovations Online*, 3(1): 18–24.

Weaver, K. and Olson, J. (2006). 'Understanding paradigms used for nursing research', *Journal of Advanced Nursing*, 53(4) 459–69.

Willis, J. W. (2007). *Foundations of Qualitative Research: Interpretive and critical approaches*. Thousand Oaks, CA: Sage.

Wilson, S. (2001). 'What is an indigenous research methodology?', *Canadian Journal of Native Education*, 25(2): 175–9.

3 Choosing an appropriate research methodology and method

Alex Opoku, Vian Ahmed and Julius Akotia

The choice of appropriate research methodology is one of the most difficult and confusing decisions for most researchers. The type of research will dictate the right research methodologies that should underpin the research and data-collection methods to be used. Regardless of the method or methodology adopted for the study, the data-collection techniques employed must be suitable and capable of meeting the objectives of the study. Moreover, it is important that the technique used to collect data is adequate to provide the information required to accomplish the overall goals of the study. This chapter builds on Chapter 2 to discuss the available research methodologies and the basis of selecting the most suitable. The chapter presents a review of relevant current literature on the choice of appropriate research methodology, sampling procedures and data-collection techniques. It highlights the strengths and weaknesses of each type of research methodology: qualitative, quantitative and mixed methods.

3.1 Introduction

There are many research strategies that can be adopted by a researcher, depending on the kind of research; such strategies include: experiment, survey, case study, action research, grounded theory, ethnography, archival research etc. (Crotty, 2004; Saunders *et al.*, 2009). The choice of research strategy is guided by the research question(s), research objectives, amount of existing knowledge, available time and resources and, finally, the philosophical underpinnings of the researcher (Saunders *et al.*, 2009). The choice of suitable data-collection and -analysis methods (qualitative, quantitative or mixed methods) for any research study is determined by the paradigm of the research and the nature of the research question. Creswell (2003) argues that no particular method has an advantage over the others, and that the actual research questions of the study should help determine the choice of method. The research methodology focuses on the process or steps and the kinds of research tool and procedure needed to obtain the required data for the study (Mouton, 2001). Generally, a research approach, whether a quantitative, qualitative or mixed method, is designed with the primary objective to collect data considered to be representative of the larger population (Gray, 2006). As it is usually not practical for the researcher to collect all the information required from the entire population, use of a sampling technique allows the assembling of such information from a segment of the population (Saunders *et al.*, 2009).

A significant number of social science studies, including construction management, involve acquiring information from the subject or topic under investigation through the use of questionnaire surveys, interviews, participant observations, etc., to fulfil their objectives. However, among these data-collection techniques, the questionnaire has been found to be the most prominent instrument used by many researchers to seek answers to research questions for their studies (Fellows and Liu, 2003). Interviews are considered to be the best data-collection option in situations where the objective of the research is concerned with the exploration of the feelings and attitudes of participants, in an attempt to gain a deeper appreciation and greater understanding of a particular phenomenon (Gray, 2006; Denzin and Lincoln, 2008). Fellow and Liu (2003) argue that interviews are useful tools to obtain detailed information about personal feelings, perceptions and opinions. The second section of the chapter discusses the various types of research methodology, and Section 3.3 describes populations and samples. Section 3.4 presents data-collection techniques, followed by illustration of the process of choosing an appropriate research method in Section 3.5. The last section highlights the conclusions and summary of the chapter.

3.2 Research methodology

The choice of a research methodology is a difficult one and should be based on the researcher's convictions, beliefs and interests (Goulding, 2002). Other important factors involved in choosing a research methodology include the aim of the research, epistemological concerns and norms of practice of the researcher and other previous work in this topic area (Buchanan and Bryman, 2007). A *methodology* refers to the philosophy and framework that are fundamentally related to the entire process of the research. Research *design* involves the plan of action that links the philosophical assumptions to specific methods (Creswell, 2003; Crotty, 2004), and *methods* are the specific techniques of data collection and analysis (Creswell, 2003). Mackenzie and Knipe (2006) argue that *methodology* is the overall approach to the proposed research linked to the paradigm or theoretical framework to be used, whereas the *method* refers to systematic modes, procedures or tools used for the collection and analysis of data. Three main methods are currently accepted for conducting research: quantitative, qualitative and mixed methods. A qualitative methodology focuses on process and meanings and uses techniques such as in-depth interviews, focus groups and participant observation (Sale *et al.*, 2002). Quantitative research, however, involves a systematic scientific investigation of quantitative phenomena and their relationships by employing mathematical models to test theories and hypotheses pertaining to the natural world (Fellows and Liu, 2003; Creswell, 2009). The use of the mixed method approach allows researchers to adopt multiple approaches to data collection and analysis in a single study.

Qualitative research method

A qualitative research method helps to address questions that cannot be answered by way of quantification (Ospina, 2004). In the qualitative research paradigm, the most important focus is for researchers to capture accurately the existing experiences and perceptions of participants involved in the phenomenon or process under investigation (Onwuegbuzie and Johnson, 2006). It is better for obtaining important preliminary insights about the phenomena being studied than quantitative surveys. Liamputtong and Ezzy (2005) add that exploratory qualitative research helps researchers to acquire information about research issues where little is known. Qualitative methods are concerned with words and images, which the researcher employs in seeking to interpret meanings and explanations of the way people behave and to develop an understanding of social constructs. The principal advantage of the qualitative method is its ability to generate very rich, deep data. Ospina (2004) summarised the benefits of using qualitative research methods as follows. They can be used to:

- add more detail to existing knowledge of a phenomenon obtained from a quantitative study;
- better understand a topic by studying it simultaneously or concurrently;
- explore a phenomenon that has not been studied before;
- develop understanding of any phenomenon in its complexity;
- help understand intricate phenomena that are difficult to approach quantitatively;
- advance a phenomenon well studied quantitatively but not well understood in any depth.

Quantitative research method

Many such discussions in favour of quantitative research within the built environment have centred on such reliability, validity and the scientific principles considered as a major strength associated with this research methodology. Creswell (2003) described quantitative research as *objective* in nature, whereas Naoum (2013) defined it as an enquiry into social or human problems based on testing a hypothesis or a theory composed of variables, measured with numbers and analysed with statistical procedures. Quantitative methods centre on objectivity and endeavour to capture reality. However, it has been asserted that quantitative methods are inflexible, myopic, mechanistic and limited to the realm of testing existing theories (Toor and Ofori, 2008). Some of the commonly used quantitative research methods are structured interviews, structured surveys, symbolic models and physical experimentation (Naoum, 2013). Quantitative methods are used in a systematic, controlled, empirical way for a significant investigation of facts about natural phenomena. However, the use of a quantitative technique stipulates that complete objectivity, as implied in the positivist position, is not possible, as all observations are driven by pre-existing theories or concepts that determine how and why objects are constituted and selected (Seale, 2002). A summary of the key characteristics of qualitative and quantitative research is presented in Table 3.1.

Table 3.1 Key characteristics of qualitative and quantitative research

Qualitative research	Quantitative research
Uses inductive approach	Uses deductive approach
Involves theory building	Involves theory testing
Employs subjective approach	Employs objective approach
Open and flexible approach	Closed and planned approach
Researcher is close to the respondents	Researcher is distant from respondents
Employs theoretical sampling	Employs random sampling
Uses explicative data analysis	Uses reductive data analysis
Low level of measurement	High level of measurement

Source: Adapted from Sarantakos, 1998

Mixed methods research

Mixed methods research is increasingly being accepted as the third major research approach and has become popular in a number of disciplines (Johnson *et al.*, 2007; De Silva, 2009). According to Teddlie and Tashakkori (2009), a mixed methods research allows the researcher to answer quantitative and qualitative questions simultaneously. A mixed method study combines elements of qualitative and quantitative research approaches (qualitative and quantitative viewpoints, data collection and analysis techniques) in a single study, concurrently or sequentially (Johnson *et al.*, 2007; Creswell *et al.*, 2008; Borrego *et al.*, 2009). Mixed methods research attempts to consider multiple viewpoints, perspectives, positions and standpoints using both qualitative and quantitative research (Johnson *et al.*, 2007). Mixed methods research aims to draw from the strengths and minimise the weaknesses of both in single research studies and across studies, but not to replace either of these approaches (Johnson and Onwuegbuzie, 2004). Although the use of qualitative studies provides eloquent, in-depth insights through subjective interpretations of experiences, adopting mixed methods allows researchers to minimise the over-dependence on statistical data to explain a social occurrence and experiences that are mostly subjective in nature (Jogulu and Pansiri, 2011). Creswell (2003) identified that using mixed methods research provides strengths that offset the weaknesses of both quantitative and qualitative research. It also provides more comprehensive evidence for studying a research problem than either using quantitative or qualitative research alone. According to Creswell (2003), a mixed methods approach helps answer questions that cannot be answered by qualitative or quantitative methodologies alone. It encourages researchers to collaborate across the sometimes adversarial relationship between quantitative and qualitative researchers and also encourages the use of multiple worldviews or paradigms. The key strengths and weaknesses of using mixed methods research methodology are presented in Table 3.2.

However, some researchers believe that mixed methods are incompatible, arguing that qualitative and quantitative methodologies are drawn from different

Table 3.2 The strengths and weaknesses of using mixed methods research

Strengths	Weaknesses
• Provides strong evidence for conclusions • Increases the ability to generalise the results • Produces more complete knowledge necessary to inform theory and practice • Answers a broader range of research questions • Uses the strength of one method to overcome the weaknesses in another method	• More expensive and time consuming • Researchers need to understand fully how to use multiple methods and approaches • Difficult when used in a single study • Can be difficult for a single researcher, especially when the two approaches are used concurrently

Source: Teddlie and Tashakkori, 2009; Creswell, 2003; Johnson and Onwuegbuzie, 2004; Bazeley, 2004

epistemological assumptions and have different research cultures that work against the merging of research methodologies (Sale *et al.*, 2002; Brannen, 2005; Scott and Briggs, 2009). In addition, the use of mixed methods research presents a number of challenges to the researcher, including the need to develop new skills, more time required to complete the research, logistical issues in conducting the research, the need to demonstrate the rigour of the additional data and the integration of research findings. Bryman *et al.* (2008) argue that the use of mixed methods through the combination of different data sources helps uncover different views, perceptions and experiences. There are four types of mixed method design, including *triangulation*, *embedded*, *explanatory* and *exploratory* design methods. The triangulation design method collects both quantitative and qualitative data at the same time and merges the results to generate greater reliability. The embedded design, however, collects data sequentially, but one set will be supportive of the other. The explanatory and exploratory designs are both two-phase methods, but, whereas the explanatory design collects number data and then narrative data to explain the number data, the exploratory mixed method design collects data in reverse order (Creswell, 2012). The triangulation mixed method design is illustrated in Figure 3.1.

3.3 Population and sampling

The main motivation for every researcher conducting research is to draw sufficient information for a meaningful analysis to be carried out so that the best conclusion can be arrived at (May, 2011). However, the major challenge researchers are often confronted with when conducting such research work is how to estimate the number of respondents required to provide them with the information, as well as the processes by which sufficient information can be generated to achieve their research objectives (Sarantakos, 1998). In view of this, a sampling technique has been seen as the most suitable means by which such estimation and information can be obtained in a manner that enables them to address the requirements of their research objectives. The population is the total number of members of the group

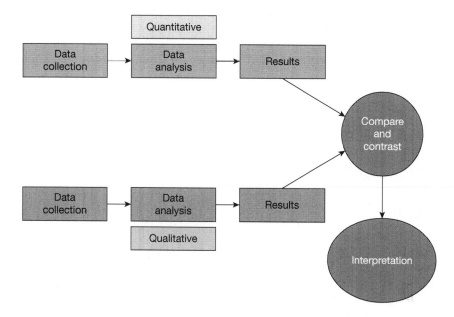

Figure 3.1 Triangulation mixed method design
Source: Adapted from Creswell, 2012

that the researcher is interested in studying, and a sample is a subset of the population that is usually chosen to serve as a representation of the views of the population. It is not practical to study the whole population, owing to restrictions on time, money and other resources (Burgess, 2001). According to Bryman (2001), the need to adopt a sampling technique is central to any research work because a sampling technique is based on sound criteria and its adoption enables researchers to estimate, identify and obtain detailed information from a reasonable number of respondents within a targeted population.

Qualitative research focuses on smaller groups of expert practitioners so as to obtain the optimum data in order to examine a particular context in great detail (Borrego *et al.*, 2009), whereas quantitative research collects data that are representative of a population and uses the information gathered to generalise findings from a drawn sample back to a population, within the limits of error (Bartlett *et al.*, 2001). In qualitative research, the guiding principle for choosing the sample size should be the concept of expertise and eventual data saturation (Mason, 2010). Table 3.3 provides a brief summary of the key differences between qualitative and quantitative sampling.

However, Naoum (2013) cautioned researchers to be careful when choosing the appropriate sample size required during the research design stage, to ensure that the sample size selected is a true reflection of the entire population. Although, usually, sample sizes are relatively small in their composition for most research

Table 3.3 A summary of the difference between qualitative and quantitative sampling

Qualitative sampling	Quantitative sampling
Relatively small sample	Relatively large sample
Cost is low	Cost is high
Less time-consuming	Time-consuming
Flexible parameters	Parameters are fixed
Occurs during data collection	Occurs before data collection
Often based on saturation	Based on probability theory
Not representative	Representative
Respondents are treated as persons	Respondents are treated as units
Sample size is not determined statistically	Sample size is determined statistically
Selection is influenced by the researcher	No researcher bias in selection

Source: Adapted from Sarantakos, 1998

projects, it is argued that careful selection may lead to a more credible and desirable outcome (Naoum, 2013). It is further argued that, irrespective of the research approach, correct estimation of sample size will enable researchers to examine the variability in the samples to draw inferences from the whole population (May, 2011). According to Sarantakos (1998, 2013) and Saunders *et al.* (2009), the estimation of the required sample size for any particular research approach should involve the consideration of issues such as: the nature of the research questions, the time and resource availability, and the characteristics of the population from which the sample is required. It is argued that, where the aim of a research is to understand common perceptions and experiences among a group of relatively homogeneous individuals, twelve interviews should be sufficient (Guest *et al.*, 2006). This study conducted an experiment on a corpus of transcripts from interviews with women in two West African countries, in which the researchers carried out a systematic analysis of transcripts of sixty interviews and found that 94 per cent of the coded topics that appeared were identified within six interviews, and saturation was attained after twelve interviews.

There are two main types of sampling technique available: these are probability or random and non-probability sampling (Sarantakos, 1998, 2013; May, 2011). Probability sampling techniques adopt well-structured, stringent procedures for the identification and selection of samples from the target populations (Sarantakos, 1998, 2013). They are useful in situations where a high degree of reliability and generalisation of the findings is required (Sarantakos, 1998). Using a probabilistic or random sampling approach also allows researchers to ensure that all participants within the defined population are proportionally represented (Black, 1999; Fisher, 2004; May, 2011). Probability sampling forms include *simple random, systematic, stratified* and *cluster*, which are generally employed for quantitative-based studies (Saunders *et al.*, 2009). Non-probability sampling techniques, in contrast, adopt approaches that are less stringent, with less emphasis on representation of samples from the larger population (Sarantakos, 1998). According to May (2011), they are mainly adopted in situations where there are no well-defined sampling frames, and yet the general features of the population are already known to the researcher. Owing to their flexible nature, they are mainly adopted by qualitative researchers when deciding which sample sizes are best suited for the study (Sarantakos, 1998). Their main forms include: *accidental, purposive, quota* and *snowball* sampling, which are usually inclined to a qualitative-based research methodology (Sarantakos, 1998; Black, 1999). One major guiding principle that determines the identification and selection of samples from the population, using either probability or non-probability types of sampling technique, is the application of a sampling frame (May, 2011; Naoum, 2013). Saunders *et al.* (2009) described a sampling frame as a complete list of all respondents located within a larger population, from which research samples are drawn. Without such an appropriate sampling frame, in which the population can be properly defined and estimated, it is impracticable for the researcher to collect a representative sample to arrive at a definitive conclusion, generalisable to the entire population (Saunders *et al.*, 2009). Researchers will then be able to generate a sample size that can generally be used to estimate the saturation points in qualitative terms and also examine the sample size statistically in quantitative terms (Sarantakos, 1998). Figure 3.2 illustrates the sampling approach when the mixed method approach is used; the sampling technique adopted is a combination of probability and non-probability sampling. A purposive sampling

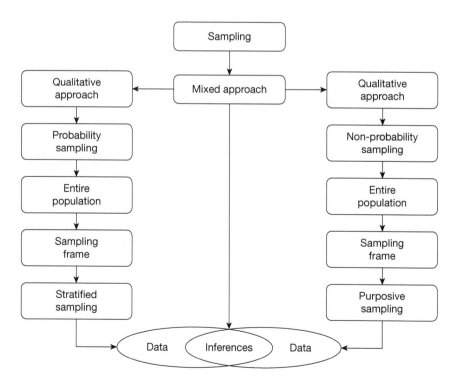

Figure 3.2 The sampling approach adopted from Saunders *et al.* (2009)

technique is used for the qualitative part of the study, whereas, on the other hand, a stratified random sampling approach is used for the quantitative study.

3.4 Data-collection techniques

There are a number of research data-collection methods for research design and these include: *focus group discussions, questionnaires, interviews* (structured, unstructured and semi-structured), record reviews (including literature) and observations (Denscombe, 2007). A major determinant of the data-collection technique is the nature and type of the enquiry and the information required about a particular setting or context (Naoum, 2007). The fundamental rationale for collecting data is to allow the researcher to gather enough evidence and, consequently, draw the inferences required to make important decisions about the findings (Tashakkori and Teddlie, 2010). Different data-collection techniques may be suitable to different research methodologies and enquiries (Pope *et al.*, 2002). However, deciding on the types of data-collection technique to adopt will depend largely on the research methodology and the overall objectives of the study (Fellows and Liu, 2003; Naoum, 2007).

Therefore, to ensure such fitness for purpose for a particular study, Naoum (2007) suggested that data-collection techniques such as personal interviews could be combined with a questionnaire survey to best understand participants' behaviour. Data-collection techniques can also be used independently or in combination, depending on the circumstances and the researcher's own judgement as to which technique(s) is best suited to obtain the required data for the study (Naoum, 2007; Saunders *et al.*, 2009). It is argued that data collected from multiple sources could complement each other to offer a more comprehensive picture for the study (Bazeley, 2004). For this reason, the adopting of multiple data-collection techniques, such as interviews, observation, a questionnaire survey and so on, provides the medium to collect both open- and closed-ended data, as required (Saunders *et al.*, 2009). Additionally, it is expected that the outcomes of such multiple data-collection approaches will yield more powerful research results than if there were just one data-collection approach (Chen, 2006). Equally, the application of multiple data-collection approaches would generally help to corroborate, complement and authenticate evidences obtained from other sources (Johnson *et al.*, 2007).

Although a distinction is commonly drawn between data-collection techniques for qualitative and quantitative research methodologies, it has been argued that the techniques can be combined in practice. It is acknowledged that using qualitative and quantitative data sources can be complementary (Saunders *et al.*, 2009). Using such an approach will enable researchers to triangulate their findings to provide more solid evidence and a better representation of the social world. For example, data collected through semi-structured interviews may be used to complement and triangulate findings obtained from questionnaire survey data. Table 3.4 illustrates the research philosophy/paradigm, the primary research methods and the suitable data-collection tools that are generally available. It shows how research methods cross philosophical/paradigm boundaries.

Table 3.4 Research paradigm, primary methods and data-collection tools

Philosophy/paradigm	Primary methods	Data-collection tools
Positivist/postpositivist	Quantitative methods	Experiments Quasi-experiments Tests Scales
Interpretivist/constructivist	Qualitative methods	Interviews Observations Document reviews Visual data analysis
Pragmatic	Qualitative and/or quantitative methods	Techniques from both positivist and interpretivist paradigms, such as interviews, observations and testing and experiments

Source: Adapted from Mackenzie and Knipe, 2006

Table 3.4 indicates that the choice of appropriate data-collection and -analysis method for a study is determined by the research philosophical stance or paradigm and the research question being considered. It is, therefore, important for a researcher to choose the right data-collection and -analysis methods for each particular study.

Questionnaires

An important part of any good research design involves making sure that the questionnaire design addresses the needs of the research questions (Burgess, 2001). Some of the advantages of using a questionnaire as a data-collection technique include flexibility, which allows it to be adapted in a diversity of theoretical positions and research questions, its relative cheapness and speed or ease of administration (Bryman, 2011). According to Saunders *et al.* (2009) and Bryman (2008), the questionnaire constitutes the most commonly used survey tool for eliciting data from a large geographical area for many research works, in comparison with the use of non-standardised data-collection techniques. It is believed that the internal validity and reliability of the findings will be enhanced to a large extent, if questionnaires are properly designed, structured, worded and administered (Naoum, 2007; Saunders *et al.*, 2009). The basic process of survey research is illustrated in Figure 3.3.

Questionnaires can be delivered to the respondent by various means, including post and email attachments, or by publishing on a website for interactive completion (Burgess, 2001). The traditional form of questionnaire survey is the postal questionnaire, but the use of electronically mailed questionnaires rather than posted questionnaires is gaining momentum owing to the increased speed and lower cost (Naoum, 2013). The use of email or Internet-based questionnaires offers more

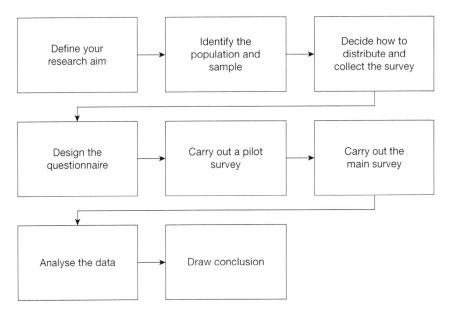

Figure 3.3 Questionnaire design process
Source: Adapted from Burgess, 2001

benefits than the traditional, posted surveys. They are much cheaper and offer significant time-saving in terms of ease of data administration. According to Bryman (2008), use of online questionnaires provides many advantages: they facilitate easy and speedy responses and reach out to a large number of respondents in a cost-effective manner, irrespective of distance and location. Similarly, by using an online service, the researcher is able to create his/her own questions speedily with the available survey-template software (Creswell, 2009), which allows for easy downloading of responses from the survey software database.

Generally, questionnaires are usually designed in two main forms: open-ended or unrestricted questions and closed-ended or restricted questions (Fellows and Liu, 2003; Naoum, 2007). The open-ended questions provide flexibility, as the respondents can respond to questions in their own way without being restricted to the researcher's line of thought. No options or predefined categories are suggested. The questions are designed to elicit full information from the respondents in an open and flexible manner. They allow respondents to provide their own answers without being constrained by a fixed set of possible answers, and they can also provide alternative answers to the problems/questions when they want to do so. Closed-ended questions, in contrast, are designed to elicit a limited set of specific responses from the respondents (Fellows and Liu, 2003). The closed-ended questions usually require straightforward answers from respondents, limited to a fixed set from which they can choose. They are usually characterised by short questions, which often require short and direct responses (Naoum, 2007), which are readily analysed by the researcher. They are useful in obtaining specific data to confirm a fact or opinion from respondents (Saunders *et al.*, 2009).

A questionnaire allows large populations to be surveyed more efficiently than other instruments, such as interviews (Denscombe, 2007), because the data can be collected simultaneously from a sample, thus providing a snapshot of its characteristics and opinions at a moment in time. The response rate could be very high if it is administered properly, especially when respondents are sent reminders. Finally, respondents get the opportunity to consult or carry out research before answering the questions. However, the disadvantages of using questionnaires include: industry fatigue, accuracy, lack of control over respondents, the requirements for simple questions that are easily understood and lack of flexibility (Fellows and Liu, 2003). The absence of the researcher means they have no opportunity to probe the issues further for respondents to elaborate more (Bryman, 2008) or to clarify any ambiguity or deviation from the issues (Naoum, 2007).

Interviews

Interviews are major data-collection techniques commonly used to elicit data, mainly for qualitative-based studies (Bryman, 2001). As they allow for social interaction and free flow of communication between the interviewer and the interviewee, they are largely regarded by many researchers as the most effective tool for gathering information that is concerned with the narration of interviewees' opinions and experiences (Qu and Dumay, 2011). An interview has been defined by Sheppard (2004) as a conversation with a purpose that allows the researcher to gain insight into the interviewees' account of their experience, perceptions and circumstances in relation to the specified research questions the interview is addressing. In making a more plausible case for an interview approach, Gray (2006) indicated that the adoption of interviews becomes necessary in the following situations, where:

- there is a real need for the researcher to obtain greater personalised information;
- there is a need for adequate probing of issues;
- a good response rate is required;
- the respondents have difficulty with writing.

According to Qu and Dumay (2011), the application of interviews provides a powerful means to discover new knowledge and capture the accounts of experts in the field in a more open, consistent and systematic manner, which the standardised methods, such as questionnaires, are unable to do. Unlike a questionnaire approach, where the objective is to obtain definite responses from a large sample, personal interviews essentially seek to obtain rich, in-depth information from interviewees within a well-controlled setting (Naoum, 2007). Fundamentally, most qualitative interviews are conducted on a face-to-face basis. Accordingly, this practice offers enormous opportunity for both parties to engage effectively and talk through the issues freely, in greater detail, without any doubt or ambiguity. Such engagement allows interviewers a great deal of latitude to probe various aspects of the issues at hand (Naoum, 2007; Denzin and Lincoln, 2008). Saunders *et al.* (2009) argued that a personal interview approach involving such one-to-one interactions can also be beneficial to researchers. One advantage is that they provide an opportunity for researchers to take a record of the interviewees' non-verbal communications (Saunders *et al.*, 2009). They also help to create a platform for researchers to explain the purpose of the study. This also allows follow-up questions to be asked to get interviewees to expatiate further on their responses. Interviews have also been criticised for their lack of a standardised approach often adopted to elicit information, which in some instances leads to a lack of rigour and reliability in the findings (Saunders *et al.*, 2009). It is also argued that data collected through such an interview approach may take some time for the researcher to transcribe and, in some cases, may be difficult to code and analyse, especially when a large number of interviewees is involved (Gray, 2006). Again, inadequate probing and long conversations on issues during the interview process can lead to insufficient and superficial responses from interviewees (Castro *et al.*, 2010).

Various forms of interview technique are available for conducting a social science research; however, the choice of any particular type must be grounded on the nature of the research questions, as well as the objectives set out for the study (Gray, 2006; Saunders *et al.*, 2009). The most commonly used ones are: *structured*, *semi-structured* and *unstructured* (Naoum, 2007). Recognising the purpose of each of these interview types, in Berg's (2007) view, forms the basis for starting the data-collection process using interviews as the main data-collection tool for the study. A commonality associated with their objectives is to obtain primary information from interviewees (Bryman, 2001). However, a major distinction identified with these forms of interview, as explained below, is their degree of rigidity in relation to their mode of presentation (Berg, 2007). Table 3.5 presents the characteristics of the structured, semi-structured and unstructured interviews.

Table 3.5 The characteristics of interview types

Structured interview	Semi-structured interview	Unstructured interview
Mainly for quantitative data	Mainly for qualitative data	Mainly for qualitative data
Captures data speedily	Captures data slowly and is time-consuming	Captures data slowly and is time-consuming
Uses random sampling	Uses purposive sampling	Uses purposive sampling
Uses strict interview format	Uses flexible interview format or schedule	Uses flexible interview format or schedule
Data usually easy to analyse	Data may sometimes be difficult to analyse	Data usually difficult to analyse
Tends to positivist view of knowledge	Mixture of positivist and interpretivist view of knowledge	Mixture of positivist and interpretivist view of knowledge

Source: Adapted from Gray, 2006

3.5 Illustrating the process of choosing appropriate research method

Table 3.6 illustrates the process of choosing an appropriate research methodology using a topic example.

Table 3.6 Illustrating the process of choosing appropriate research methodology

---

Research topic: The role of organisational leadership in promoting sustainable construction practices in construction organisations

---

Research aim: To investigate the link between leadership and sustainability and produce a support framework towards the promotion of sustainability practices in the delivery of construction projects through leadership within construction organisations

---

Research process	*Selected research tools and procedures*	*Reasons for the choice*
The philosophy of the research	*Pragmatic philosophical stance*: pragmatic research adopts both positivist and interpretivist paradigms using quantitative and qualitative methods to match specific research questions	Leadership and sustainability are best understood through diverse theoretical positions, research methods and the examination of a great variety of research contexts and settings
Research approach	*Abductive reasoning* (logic of the research): abductive reasoning takes the middle ground between deductive (theory testing) and inductive (theory building) involving a piecemeal set of observations and proceeds to the best possible solution with information at hand	Leadership and sustainability have many dimensions, and, therefore, it is asserted that different research approaches must be used to reveal relevant data for each research question under exploration
Research strategy	*Multiple strategies*: literature review, interviews and survey	No single strategy can solve the leadership and sustainability research problem, hence the use of multiple strategies
Research choices or methodology	*Mixed method*: a mixed method research approach combining both qualitative and quantitative methods helps in carrying out an in-depth study of the 'leadership and sustainability' phenomenon	A mixed methods research approach provides both qualitative and quantitative evidence, which gives a more complete picture of the link between leadership and sustainability in the UK construction industry
Time horizon (if applicable)	*Cross-sectional*: time is predetermined for data collection, data analysis and other research tasks	The time horizon adopted for the topic is cross-sectional and has to be completed within a specified timeframe, like any academic research

continued . . .

Table 3.6 Continued

Research process	Selected research tools and procedures	Reasons for the choice
Method(s) used for data collection	*Questionnaire and interview:* because of the broad scope and the context of this study (leadership and sustainability), a mixed methods approach for data collection, using both interview and questionnaire research techniques, was adopted to achieve the research aim and objectives	The interview provides in-depth understanding of the state of the art in practice, and the questionnaire survey provides a wider view of leaders in construction organisations with the role of leading the change towards the adoption of sustainability practices in construction project delivery
Unit of study	The unit of analysis in this study is *leadership within contractor and consultant organisations* in the UK construction industry	To establish organisational leadership role in promoting sustainability drivers, challenges, roles, styles and engaged sustainability practices
Data analysis technique	The questionnaire data collected are analysed using *Statistical Package for Social Sciences (SPSS)* and Excel, and data from the interviews are analysed with *Nvivo data management software*	The SPSS will be used to examine any cross-tabulation or associations or grouping that emerge from the survey data through factor and coding analysis, and the qualitative Nvivo software will be used to assist coding and derivation of themes from the interview data

3.6 Conclusions

This chapter presents research methodologies and procedures that can be adopted in order to achieve the research aim and objectives. It highlights the advantages and disadvantages, and the strengths and weaknesses, of the three research methodologies: qualitative, quantitative and mixed method. Whereas the traditional characteristics of quantitative enquiry are based on deductive reasoning, statistical analysis and hypothesis testing, the traditional characteristics of qualitative enquiry, on the other hand, are based on inductive reasoning and hypothesis generation. Given that construction processes are fundamentally complex, with diverse players and rapid technological changes, both deductive and inductive reasoning have crucial roles to play in ensuring the successful exploration of these issues. Combining both qualitative and quantitative methods helps in carrying out an in-depth study of the phenomenon under investigation. It is normally the practice that researchers associate small samples with qualitative research and large samples with quantitative research; however, the size of the sample should primarily be guided by the research objective, research question(s) and the research design. Although questionnaire surveys present a convenient way of gathering data much faster from larger population, they are said to be most useful and beneficial when complemented by other data-collection techniques, such as interviews.

References

Bartlett, J. E., Kotrlik, J. W. and Higgins, C. C. (2001). 'Organizational research: Determining appropriate sample size in survey research', *Information Technology, Learning, & Performance Journal*, 19(1): 43–50.

Bazeley, P. (2004). Issues in mixing qualitative and quantitative approaches to research. In R. Buber, J. Gadner and L. Richards (eds), *Applying Qualitative Methods to Marketing Management Research* (pp. 141–56). Basingstoke, UK: Palgrave Macmillan.

Berg, B. L. (2007). *Qualitative Research Methods for the Social Sciences* (6th edn). San Francisco, CA: Pearson Education.

Black, T. (1999). *Doing Quantitative Research in the Social Sciences: An integrated approach to research design, measurement and statistics.* London: Sage.

Borrego, M., Douglas, E. P. and Amelink, C. T. (2009). 'Quantitative, qualitative, and mixed research methods in engineering education', *Journal of Engineering Education*, 98(1): 53–66.

Brannen, J. (2005). Mixed Methods Research: A discussion paper, NCRM Methods Review Papers, ESRC National Centre for Research Methods, Southampton, UK.

Bryman, A. (2001). *Social Research Methods.* New York: Oxford University Press.

Bryman, A. (2008). *Social Research Methods* (3rd edn). Oxford, UK: Oxford University Press.

Bryman, A. (2011). 'Mission accomplished? Research methods in the first five years of leadership', *Leadership*, 7(1): 73–83.

Bryman, A., Becker, S. and Sempik, J. (2008). 'Quality criteria for quantitative, qualitative and mixed methods research: A view from social policy', *International Journal of Social Research Methodology*, 11(4): 261–76.

Buchanan, D. and Bryman, A. (2007). 'Contextualizing methods choice in organizational research', *Organizational Research Methods*, 10: 483–501.

Burgess, F. T. (2001). *A General Introduction to the Design of Questionnaires for Survey Research: Guide to the design of questionnaires* (1st edn). Leeds, UK: University of Leeds.

Castro, F. G., Kellison, J. G., Boyd, S. J. and Kopak, A. (2010). 'A methodology for conducting integrative mixed methods research and data analyses', *Journal of Mixed Methods Research*, 4(4): 342–60.

Chen, H. T. (2006). 'A theory-driven evaluation perspective on mixed methods research', *Research in the Schools*, 13(1): 75–83.

Creswell, J. W. (2003). *Research Design: Qualitative, quantitative, and mixed methods approach* (2nd edn). Thousand Oaks, CA: Sage.

Creswell, J. W. (2009). *Research Design: Qualitative, quantitative and mixed methods approaches* (3rd edn). Thousand Oaks, CA: Sage.

Creswell, J. W. (2012). *Educational Research: Planning, conducting, and evaluating quantitative and qualitative research* (4th edn). Boston, MA: Pearson.

Creswell, J. W., Plano Clark, V. L. and Garrett, A. L. (2008). Methodological issues in conducting mixed methods research designs. In M. M. Bergman (ed.), *Advances in Mixed Methods Research: Theories and applications* (pp. 66–83). Thousand Oaks, CA: Sage.

Crotty, M. (2004). *The Foundation of Social Research: Meaning and perspective in the research process.* London: Sage.

Denscombe, M. (2007). *The Good Research Guide* (3rd edn). Maidenhead, UK: Open University Press.

Denzin, N. K. and Lincoln, Y. S. (2008). *The Landscape of Qualitative Research* (3rd edn). London: Sage.

De Silva, T. (2009). *Benefits of Mixed Methods in Environmental Reporting Research.* Christchurch, NZ: Lincoln University.

Fellows, R. and Liu, A. (2003). *Research Methods for Construction Students* (2nd edn). Oxford, UK: Blackwell.

Fisher, C. (2004). *Researching and Writing a Dissertation for Business Students*. Harlow, UK: Pearson Education.

Goulding, C. (2002). *Grounded Theory: A practical guide for management, business and market researchers*. London: Sage.

Gray, D. E. (2006). *Doing Research in the Real World*. London: Sage.

Guest, G., Bunce, A. and Johnson, L. (2006). 'How many interviews are enough? An experiment with data saturation and variability', *Field Methods*, 18(1): 59–82.

Jogulu, U. D. and Pansiri, J. (2011). 'Mixed methods: A research design for management doctoral dissertations', *Management Research Review*, 34(6): 687–701.

Johnson, B. R. and Onwuegbuzie, A. J. (2004). 'Mixed methods research: A research paradigm whose time has come', *Educational Researcher*, 33(7): 14–26.

Johnson, B. R., Onwuegbuzie, A. J. and Turner, L. A. (2007). 'Toward a definition of mixed methods research', *Journal of Mixed Methods Research*, 1(2): 112–33.

Liamputtong, P. and Ezzy, D. (2005). *Qualitative Research Methods*. South Melbourne, VIC: Oxford University Press.

Mackenzie, N. and Knipe, S. (2006). 'Research dilemmas: Paradigms, methods and methodology', *Issues in Educational Research*, 16(2): 193–205.

Mason, M. (2010). 'Sample size and saturation in PhD studies using qualitative interviews', *Forum: Qualitative Social Research* [online journal], 11(3): Article 8. Available at www.qualitative-research.net (accessed 21 October 2010).

May, T. (2011). *Social Research: Issues, methods and process* (4th edn). Maidenhead, UK: McGraw-Hill Education.

Mouton, J. (2001). *How to Succeed in Your Master's and Doctoral Studies, A South African Guide and Resource Book*. Pretoria: Van Schaik.

Naoum, S. G. (2007). *Dissertation Research and Writing for Construction Students* (2nd edn). Oxford, UK: Butterworth-Heinemann.

Naoum, S. G. (2013). *Dissertation Research and Writing for Construction Students* (3rd edn). London: Routledge.

Onwuegbuzie, A. J. and Johnson, R. B. (2006). 'The validity issue in mixed research', *Research in the Schools*, 13(1): 48–63.

Ospina, S. (2004). Qualitative research. In G. Goethals, G. Sorenson and J. MacGregor (eds), *Encyclopaedia of Leadership* (pp. 1280–5). London: Sage.

Pope, C., Royen, P. V. and Baker, R. (2002). 'Qualitative methods in research on healthcare quality', *Quality Safe Health Care*, 11: 148–52.

Qu, S. Q. and Dumay, J. (2011). 'The qualitative research interview', *Qualitative Research in Accounting & Management*, 8(3): 238–64.

Sale, J. E. M., Lohfeld, L. H. and Brazil, K. (2002). 'Revisiting the quantitative–qualitative debate: Implications for mixed methods research', *Quality & Quantity*, 36: 43–53.

Sarantakos, S. (1998). *Social Research* (2nd edn). Melbourne, VIC: Macmillan Education.

Sarantakos, S. (2013). *Social Research* (4th edn). Basingstoke, UK: Macmillan.

Saunders, M., Lewis, P. and Thornhill, A. (2009). *Research Methods for Business Students* (5th edn). Harlow, UK: Pearson Education.

Scott, P. J. and Briggs, J. S. (2009). 'A pragmatists argument for mixed methodology in medical informatics', *Journal of Mixed Methods Research*, 3(3): 223–41.

Seale, C. F. (2002). Computer assisted analysis of qualitative data. In F. Gubriumand and J. A. Holstein (eds), *Handbook of Interview Research* (pp. 651–70). Thousand Oaks, CA: Sage.

Sheppard, M. (2004). *Appraising and Using Social Research in the Human Services: An introduction for social work and health professionals*. London: Jessica Kingsley.

Tashakkori, A. and Teddlie, C. (2010). 'Putting the human back in "human research methodology": The researcher in mixed methods research', *Journal of Mixed Methods Research*, 4(4): 271–7.

Teddlie, C. and Tashakkori, A. (2009). *Foundations of Mixed Methods Research: Integrating quantitative and qualitative approaches in the social and behavioural sciences*. Thousand Oaks, CA: Sage.

Toor, S. R. and Ofori, G. (2008). Grounded theory as an appropriate methodology for leadership research in construction. In R. Haig and D. Amaratunga (eds), *Proceedings of the CIB International Conference on Building Education and Research: Building Resilience* (pp. 1816–31). Kandalama, Sri Lanka, 11–15 February.

Part II

Quantitative research

4 A questionnaire survey of building surveying education

The graduate voice

Mike Hoxley

A quantitative study involving a questionnaire is perceived by many researchers to be a difficult option because of increasingly high non-participation and the need to undertake detailed statistical analysis. This case study demonstrates that, with careful design, it is possible to encourage participation from a sample who can see the relevance of research to their own situation. The research approach, design of the research instrument and the data-collection and -analysis stages of the case study are all considered in some detail. The problem investigated was the current state of building surveying (BS) education in the UK. The relatively high referral rate of the main BS qualification route, shortages of graduates and disquiet expressed by employers over the content of academic courses have led many commentators to question the fitness for purpose of BS degrees. Previous studies of this subject have concentrated on talking to higher-education (HE) providers and large employers, but this case study has elicited the views of those who have graduated recently and moved into industry. An online survey was completed by more than 800 graduates, and their preparation for professional training and the main types of work undertaken by practitioners were investigated. An incentive to participate in the survey included the offer of early feedback on the results of the study. The main findings of this study are that graduates believe that more technical subjects are of greatest use to them in the workplace, and there are some interesting differences of emphasis between graduates from undergraduate and postgraduate courses.

4.1 Introduction

This quantitative case study investigated graduates' opinions of the BS education they received. An online questionnaire was used to elicit responses from more than 800 graduates. Within a few decades, the profession of BS has grown from a branch of general practice surveying to a professional group of 28,000 members worldwide – growth that has been described as 'rapid and spectacular . . . that pays tribute to the increasing importance of the building surveyor role' (Sayce, 2010). There are currently about 8,000 chartered building surveyors working in the UK, and another 700 working overseas – although nearly half of these are based in Hong Kong (RICS, 2011). There are twenty-seven Royal Institution of Chartered Surveyors (RICS)-accredited BS degree courses in the UK, and approximately one-third of graduates applying for BS posts are from postgraduate courses (Gough, 2010). In addition to a RICS-accredited degree, chartered surveyors are required to pass the Assessment of Professional Competence (APC), which involves a minimum of 2 years approved professional training and a final professional interview with two or three practitioner assessors (RICS, 2014). The first-time referral rate for the BS pathway is typically 50 per cent, which is higher than for other professional groups. Sayce (2010) suggests that the technical nature of building surveying is the likely reason that APC candidates lack the depth of experience required after 2 years in the workplace.

The high APC referral rate and general disquiet about the content of academic courses have led many in the profession to question whether BS education is fit for purpose. In a study of large BS employers, Hoxley and Wilkinson (2006) found that there were concerns about BS graduates' level of knowledge of construction technology and contract administration. The RICS HE Policy Manager for Education and Qualifications Standards, Nick Evans, has commented that there has been criticism of HE's role in supporting professional practice – principally over the quality of graduates (Evans, 2010).

As the UK emerges from deep economic recession, there are substantial fears that there will be a shortage in the supply of construction professionals, including building surveyors. There were such skills shortages in the immediate aftermath of the last two recessions, and fewer graduate places have been offered by BS firms because of the decline in activity. Sayce reports that the first generation of building surveyors is nearing retirement, and there are concerns about the supply of graduates and recently qualified surveyors to replace them. Between 2007 and 2009, BS APC enrolments fell by more than 50 per cent (Sayce, 2010). In the previous two recessions, enrolments on to HE courses in building surveying fell drastically as the number of graduate opportunities diminished, although there was a lag of several years between what happened in the jobs market and enrolment on to HE courses.

The oversight of surveying academic programmes of study by the professional body (RICS) has undergone change in the last few years. An accreditation visit, every 5 years or so, has been replaced by an annual partnership meeting. Whereas the visit would typically last a couple of days, the partnership meeting is usually over within a few hours. Employers sit on the partnership committees, and any new programmes, or changes to programmes, are considered by the committees. However, time pressures usually ensure minimal consideration and discussion of any changes in curricula. Of course, universities have their own rigorous validation and review procedures, and employers participate in these processes, but the reality is that the oversight of the curricula of surveying programmes of study has come to have a 'lighter touch' with the move to partnership meetings. It is, therefore, incumbent upon individual course providers to ensure that their curricula are current and relevant.

RICS does publish guidance on what should be included in BS courses. For many years, a document, 'Mind the Gap' (Mole, 1997), was the design template for courses. This document suggested that any profession is ultimately concerned with the way its members operate in practice and the skills, knowledge and competencies they have. The report was heavily influenced by the work of Eraut (1994), who suggested that the abilities of each professional depend on their particular knowledge and skills, as derived from their educational and practice background. Eraut (1994) considers this to be a mix of the following: propositional knowledge; personal knowledge, impressions and experiences; professional knowledge, referred to as 'process knowledge', which is based upon professional experience and action; and moral or ethical principles.

Mole's (1997) contribution has been replaced by 'Keep Learning: A Framework for Building Surveyors' (RICS, 2009). This document includes much of the philosophical discussion of the earlier 'Mind the Gap' report (Mole, 1997) and, when considering BS courses, suggests that typical modules are: construction technology, law and responsibility, economics and finance, building pathology, planning and design, environmental science and management. Murray (2010), writing in a special edition of the RICS *Building Surveying Journal* devoted to education and training, suggests that the core subjects to be covered should be:

- building pathology (in both traditional and modern methods of construction);
- environmental and material sciences;
- construction technology; and
- the laws of contract, tort, property, and landlord and tenant.

The general advice that RICS gives all universities offering programmes of study in surveying is that the APC competencies, and particularly the 'core' competencies, should be covered by the course. Obviously, in some cases, this may involve teaching the knowledge that these competencies require, even if the skills elements will be delivered as part of the later professional training. It is well known throughout industry that different BS courses have different flavours, and, indeed, some employers make a point of only appointing graduates from one or two particular universities each year. Thus, each BS programme of study will be different, but what is important, of course, is that each one delivers the core subject knowledge and understanding to prepare graduates for the workplace.

In the UK, the Higher Education Council for England uses an independent national student survey (HEFCE, 2014) as one of the proxies of quality. As far as surveying is concerned, Lee and Hogg (2009) took a similar approach when considering the education and training of the quantity surveying (QS) profession. They report the outcomes of a survey of 425 early-career QSs to rate their own degree of confidence when performing a number of standard QS tasks. Poon *et al.* (2011) took a similar approach in their study of UK real-estate graduate competencies. In their study of previous RICS education reform on the BS profession, Wilkinson and Hoxley (2005) surveyed course providers and large employers (Hoxley and Wilkinson, 2006), but the individuals who are likely to have the most clearly focused views about education are those who have recently experienced it and who are now working in industry. A decision was taken, therefore, to carry out a survey of BS graduates. The design of the research instrument, detailed accounts of the data-collection and -analysis phases and the main findings of this case study follow.

4.2 Research methodology

The aim of this study meant that the views of BS graduates needed to be elicited. A qualitative study involving interviewing graduates was considered, but this approach was rejected on the grounds that it would be difficult to secure a sample size that would enable statistically significant conclusions to be drawn. An early decision was made, therefore, to adopt a positivist approach using a questionnaire survey. Of course, very few studies are entirely deductive or entirely inductive. There was a qualitative element to the survey, in that the views of participants on the subject under investigation were sought, but the study mainly took a positivist stance, using a questionnaire to survey participants, and the data collected were subjected to statistical analysis. The study was cross-sectional in that it collected data at a single point in time, rather than over a longer period. However, participants were required to reflect on their past experiences in order to complete the survey.

The networking benefits of attending academic conferences are well known: for example, doctoral students frequently first meet their external examiner at such events. What is less often acknowledged is the potential of conferences to provide valuable insights into potential new research projects or alternative approaches to collecting data. The research reported in this chapter had its origins in a presentation of the study into quantity surveying education referred to above (Lee and Hogg, 2009) at a RICS conference held in Cape Town in 2009.

In any questionnaire study, the data-collection and data-analysis stages are relatively straightforward; what takes the time is the design of the study, and it is on this stage that the first part of this chapter will concentrate. Questionnaires are research instruments for use in a survey setting and are intended to measure something, which may include people's attitudes. The instrument can be administered by post, in face-to-face interview, over the telephone or, increasingly, by email or directly over the web. If the phenomenon to be measured describes something – for example, the voting intentions of the electorate in a political poll – then the survey can be categorised as descriptive. On the other hand, if the measurement is looking for association or causality – for example, the effect of procurement route on profitability – then the survey is said to be analytical (Oppenheim, 1992). In reality, of course, even analytical surveys contain some descriptive variables that are necessary to define the sample and to provide the independent or predictor variables.

Each question in the survey generates a variable that is coded with (usually) a number assigned to each possible response. Although numbers are usually used for coding, it is vitally important to be aware of the type (or level of measurement) of each variable. Categories, such as 'procurement route', generate nominal variables, and, if 'profitability' in our simple example is measured in actual sums of money, then this will be an interval or continuous variable. Interval variables that have meaningful zero points are known as ratio variables. The other types of variable are ordinal, which are created by ordered categories, such as those generated by attitude scales. The reason that we need to be aware of levels of measurement is so that appropriate statistical analysis techniques can be used to analyse data.

As it was respondent attitudes to the education received that were being solicited by this study, an attitude scale was adopted. The most frequently used attitude scale was developed by Rensis Likert (Likert, 1932) and employs declarative statements and a list of response categories, typically five or seven. A five-point scale, for example, may have 'strongly agree', 'agree', 'neutral', 'disagree' and 'strongly disagree' as the responses. Note, however, that there is no reason why the responses should be balanced (Tull and Hawkins, 1984). The author has used an unbalanced scale when it was known that respondents had overwhelmingly positive views about particular statements. For this case study, the five-point, balanced Likert scale described above was adopted as the principal research instrument.

Strict interpretation of the rules of measurement requires that the data generated by Likert scales should be treated as ordinal. However, some researchers (e.g. Weisberg *et al.*, 1996) argue that, provided the intervals between the various possible responses are approximately equal, then such data can be regarded as interval data, which of course enables more sophisticated statistical techniques to be used. This is an important issue that needs to be explicitly addressed by any researcher who treats attitude data as interval data.

The principal objective of the research questionnaire was to see how graduates perceived that the course they had studied prepared them for the BS APC competencies. For each APC pathway, the competencies are divided into *mandatory* (common across all of the surveying disciplines), *core* and *optional*. One would expect all of the mandatory and core competencies to be covered to some extent by BS courses and some, but probably not all, of the optional competencies. In addition to assessing confidence in APC competencies, the original questionnaire asked graduates what the most and least useful subjects they studied were, if any subject had proved of no use to them so far in their careers, as well as questions about the acquisition of knowledge and skills and their preferred methods of learning, teaching and assessment. The questionnaire was piloted with three members of the RICS BS Professional Group Board (including the chair and university representative) to obtain their views. Following this consultation, an additional question was included to assess graduates' confidence in undertaking several standard BS activities.

The success of any questionnaire survey hinges on achieving an adequate sample size. Frequently, students concentrate on the response rate they achieve, but this in itself is unimportant except in so far as it impacts on sample size. Questionnaire fatigue is a particular problem for those considering the use of questionnaires, and this, together with the considerable efforts made to secure a good response rate for this study, will be considered in greater detail in the following section.

4.3 Data-collection strategy

The traditional way to gather survey data for research has been by postal questionnaire. When postal questionnaires are administered, great care has to be taken to ensure an adequate response rate. Typical response rates quoted in textbooks have a mean of about 30 per cent, but one has to work hard to achieve this level of response. When the author first used a postal questionnaire, more than 20 years ago, he achieved very nearly a 70 per cent return rate. His doctoral study (Hoxley, 2000) achieved a response rate of nearly 50 per cent, but, when a follow-up study was conducted some 8 years later, using almost identical instruments sent to a similar sample, a response rate of only 19 per cent was achieved (Hoxley, 2007). These examples highlight just how difficult it is to achieve a sample of adequate size using postal administration. Steps that can be taken to achieve the best possible completion rate include sending the questionnaire out with an individually addressed letter explaining the purpose and relevance of the study. This involves compiling a database of these individuals and using a 'mail merge' facility to create the letters. A stamped, addressed envelope should be sent out with each questionnaire – even though the majority of these will be thrown away! It may be necessary to send follow-up letters to non-respondents. There is evidence (Berdie *et al.*, 1986) that using a term other than 'questionnaire' also increases the response rate. One way of encouraging a response is to promise feedback on the results of the research. It goes without saying, however, that, if promised, such feedback must be provided. Clearly, the anticipated response rate needs to be factored into the decision on the number of questionnaires to be sent out in order to achieve the minimum sample size required for the study.

For any given population, there is a need to achieve a certain sample size in order to reduce the likelihood of sampling error. In addition to statistical techniques, some researchers advocate estimating sample size using tables. Thus, Sarantakos (2005: 173) provides a table based on a sampling error of 5 per cent, which suggests that, for a population of about 30,000, a sample of size 379 would be required. The difficulties in achieving an adequate sample size using postal administration, changes in technology and increased postal charges have meant that questionnaire administration over the web has largely replaced the traditional postal method.

4.4 Data-collection method

For this case study, RICS Education provided the names and email addresses of current and recent APC candidates for the BS route (the 'sampling frame'). Many hours were invested in ensuring that each of the nearly 3,000 emails was individually addressed to each member of the sampling frame, using their first name. Although this is relatively easy to achieve using the mail-merge facility in MS Word, not all of the first names were contained within the RICS spreadsheet, and it was necessary to use a separate database (RICS, 2011) to ensure that the first names were available. Several studies have suggested that using personal salutations in email requests to complete surveys has a significant impact on response rate (see, for example, Heerwegh, 2005).

Peterson (2000) offers advice on how questions should be worded, and included in this advice is the need for brevity (a long questionnaire will not be completed), unambiguous questions and *relevance*. This latter point is very important, as the sample is much more likely to respond if it perceives that the survey is of interest to it and also if early feedback on the results of the study is offered to respondents.

The questionnaire was developed using the SurveyMonkey software tool. This is very easy to use, and the research team in which the author was based at the time of the survey subscribed to the advanced version of the software. There were eight sections to the questionnaire: notes explaining the purpose of the study; respondent information; how the BS course studied prepared for the APC competencies; knowledge acquisition; skills acquisition; teaching, learning and assessment; a section where respondents could leave any additional comments; and, finally, a concluding section in which respondents were thanked and could request feedback on the results of the study.

An individually addressed email was sent to the 2,910 graduates in the sampling frame, with an invitation to complete the survey via SurveyMonkey. Two hundred and twenty-six messages were undelivered for one reason or another, and 806 useable responses were received, which represents an overall response rate of exactly 30 per cent. As discussed previously, this is a very good response rate for this type of survey, and the author believes that this is partly due to the care taken in its design, but also because the subject of their own education is very important to most professionals.

4.5 Data analysis

In addition to providing results in an MS Excel spreadsheet, SurveyMonkey has the facility to calculate descriptive statistics and also to produce very good charts. The spreadsheet of the coded responses of this study was imported into SPSS, which was used to undertake the more detailed analysis and also to draw additional charts. The declared specialisations of the sample are presented in Figure 4.1. As will be seen, when asked to describe their main area of work activity, there were approximately equal numbers specialising in project work and professional work, including condition assessment (which are the two principal areas of work undertaken by building surveyors). The sample of graduates included 650 who had studied an undergraduate qualification (and 97 per cent of these had a degree), and 129 had studied for either a masters or a graduate diploma. Just under two-thirds had studied full time, 20 per cent part time, and just under 8 per cent by distance learning. A total of 614 (76 per cent) had undertaken some form of placement or work experience during their studies. The mode of the year of graduation was 2004, and the mode of the year of qualification as a chartered surveyor was 2007. The mean time since graduation was 7 years for undergraduates and 5 years for postgraduates. One hundred and thirty-three of the graduates work in the public sector, whereas the majority (65 per cent) work in private practice.

The three most commonly used statistics for the measure of central tendency (or 'average') are the mode, median and mean. Thus, reference to the mode in the previous paragraph indicates that the dates given were the most frequently occurring responses to those particular questions. The mean (which is the arithmetical average) is for use with interval data, but, as discussed previously, many researchers believe that it is also appropriate to use the mean for ordinal data, such as those generated by these attitude scales. When the results were interpreted, the

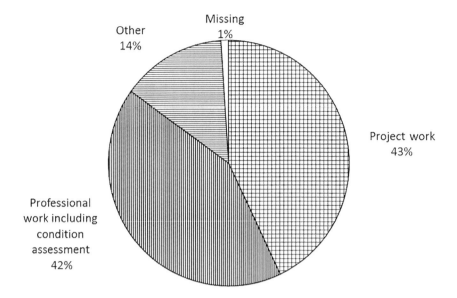

Figure 4.1 Main areas of work

'acceptable' threshold of the aggregate view of all the attitude scale questions was taken as a mean of '3' – that is, the neutral response. Above that level, the aggregate view was positive, and, below that level, it was negative. Thus, the mean of each response to each Likert-scale question was computed, and an example of the charts used to interpret the results is provided in Figure 4.2. The variables are ordered from left to right by the ranking of their means, and it will be seen that the graduates were of the view that their courses had not prepared them adequately to undertake the tasks indicated on the right-hand side of the figure.

In some branches of surveying, such as real estate, there are now more students graduating from postgraduate than undergraduate courses. As reported previously, this situation is far from the case with BS (81 per cent of this sample had studied an undergraduate course). However, the author was interested in the different views expressed by graduates from the different levels of qualification. To determine differences of emphasis between undergraduate and postgraduate responses, the non-parametric Mann–Whitney U-test was used, and the test was conducted at a 1 per cent probability level (Dancey and Reidy, 2007). Non-parametric tests are required when the data are not normally distributed. The Mann–Whitney U-test is appropriate because clearly the variables are unlikely to be normally distributed, as they have all been selected on the basis of their importance to BS education or training. In a question on skills acquisition from their course, there is a distinct divergence of view between graduates from the different levels of study. As can be seen from Figure 4.3, all of the variables have a significant difference at the 1 per cent probability level.

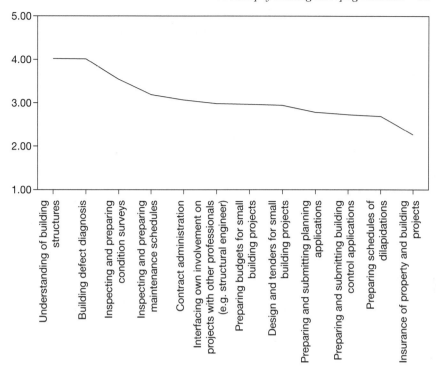

Figure 4.2 Preparation for key BS tasks

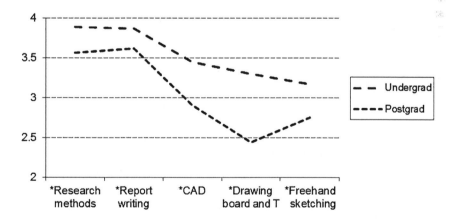

Figure 4.3 Skills acquisition

Table 4.1 Most useful and least useful subjects studied

Most useful subject studied	n	Least useful subject studied	n
Building pathology	281	Economics	170
Construction technology	231	Law	44
Law	83	Management	36
Design	30	Facilities management	33
Contract administration	29	Structures	26
Professional practice module	21	Statistics	25

When responses to the most useful and least useful subjects studied on their course were analysed, it was necessary to manually code many responses, as different subjects were given slightly different names by respondents. The top six responses to both questions can be seen in Table 4.1. Those designing programmes of study are advised to take these responses into account when deciding upon the curricula of their courses. Clearly, there are mixed messages (for example, the fact that law appears in both lists), but the overriding importance of building pathology and construction technology and the apparent irrelevance of economics are the main findings about curricula.

4.6 Main findings and conclusions

The results of this survey suggest that, of the mandatory competencies of the RICS APC, ethics, conflict avoidance, business planning, accounting and client care could be better covered on BS courses. Core competencies are adequately covered, but a high proportion of respondents believed that they should have studied (but didn't) contract administration and dilapidations. Both of these subjects are central to the work of building surveyors, and it is essential that they are adequately taught on BS courses. The optional APC competencies of insurance, risk management, work progress monitoring, financial control and commercial management of projects could be better covered on courses.

Graduates feel less than prepared to undertake the following types of work: insurance, dilapidations, submitting statutory control applications and design and tendering for small building projects. Clearly, this finding does not throw a positive light on the education that these graduates have received. Construction technology and building pathology are the most useful subjects studied, whereas economics appears to be the least useful. Although only 16 per cent of the respondents studied a postgraduate programme, there are some interesting differences of emphasis revealed, particularly in the area of skills development.

The shortcomings revealed by this survey should be addressed in future BS course designs/reviews. This mainly involves strengthening courses by ensuring adequate technical content and including core subjects, such as contract admin-istration and dilapidations. Each of these subjects is core to the work undertaken as part of the two facets of BS – project work and condition assessment. As evidenced by Figure 4.1, BS is a very broad subject, and, in the author's view, the only way adequately to address the issue of the high referral rate in the APC is to give consideration to allowing building surveyors to specialise (in professional work, including condition assessment, *or* project work) prior to the APC final assessment. Naturally, the profession has reservations about such a development, but, unless the issue of the high referral rate is addressed, potential building surveyors will be dissuaded from entering the profession in the first place.

This chapter has presented a case study of positivist research employing a questionnaire survey, the sample size of which was more than 800. The results suggest that, with careful design of the research instrument and the email introducing the survey, it is still possible to achieve an adequate sample size for a questionnaire study. Those researchers undertaking such web-based surveys are encouraged to make use of proprietary platforms such as SurveyMonkey, but they must also be aware that statistical analysis beyond that provided by such standard software will need to be carried out to achieve meaningful results.

References

Berdie, D. R., Anderson, J. F., and Niebuhr, M. A. (1986). *Questionnaires: Design and use.* London: Scarecrow Press.

Dancey, C. P. and Reidy, J. (2007). *Statistics Without Maths for Psychology Using SPSS for Windows.* Harlow, UK: Pearson Education.

Eraut, M. (1994). *Developing Professional Knowledge and Competence.* London: Falmer Press.

Evans, N. (2010). 'Collaborate to educate', *Building Surveying Journal,* July–August: 9.

Gough, H. (2010). 'Nurturing the next generation', *Building Surveying Journal,* July–August: 5.

Heerwegh, D. (2005). 'Effects of personal salutations in email invitations to participate in a web survey', *Public Opinion Quarterly,* 69(4): 588–98.

HEFCE. (2014). National Student Survey. Available at www.hefce.ac.uk/lt/nss/results/ 2015/ (accessed 2 December 2015).

Hoxley, M. (2000). 'Are fee tendering and construction professional service quality mutually exclusive?', *Construction Management & Economics,* 18(5): 599–605.

Hoxley, M. (2007). 'The fee tendering and service quality issue revisited', *Property Management,* 25 (2): 180–92.

Hoxley, M. and Wilkinson, S. (2006). 'The employers' perspective of the impact of RICS education reform on building surveying', *Structural Survey,* 24(5): 405–11.

Lee, C. and Hogg, K. (2009). 'Early career training of quantity surveying professionals', *Proceedings of Construction and Building Research (COBRA) Conference,* Cape Town, South Africa, September.

Likert, R. (1932). 'A technique for the measurement of attitudes', *Archives of Psychology,* 140: 5–55.

Mole, T. (1997). Mind the Gap: Report on the rationale and expectation of the Building Surveyors Division in the education and training of chartered surveyors. RICS, London.

Murray, J. (2010). 'Be part of the process', *Building Surveying Journal,* July–August: 12–13.

Oppenheim, A. N. (1992). *Questionnaire Design, Interviewing and Attitude Measurement.* London: Continuum.

Peterson, R. A. (2000). *Constructing Effective Questionnaires.* London: Sage.

Poon, J., Hoxley, M. and Fuchs, W. (2011). 'Real estate education: An investigation of multiple stakeholders', *Property Management,* 29(5): 468–87.

RICS. (2009). Keep Learning: A Framework for Building Surveyors. RICS, London.

RICS. (2011). 'Find a surveyor.' Available at www.ricsfirms.com/ (accessed 2 December 2015).

RICS. (2014). Building Surveying: Assessment of professional competence, building surveying pathway guide. Available at www.rics.org/uk/apc/pathway-guides/construction-pathway-guides/building-surveying/ (accessed 1 December 2015).

Sarantakos, S. (2005). *Social Research*. Basingstoke, UK: Palgrave MacMillan.

Sayce, S. (2010). 'Maintaining the supply', *Building Surveying Journal*, July–August: 6.

Tull, D. S. and Hawkins, D. I. (1984). *Marketing Research Measurement and Method*. London: Collier MacMillan.

Weisberg, H. F., Krosnick, J. A. and Bowen, B. D. (1996). *An Introduction to Survey Research, Polling and Data Analysis*. London: Sage.

Wilkinson, S. and Hoxley, M. (2005). 'The impact of RICS education reform on building surveying', *Structural Survey*, 23(5): 359–70.

5 Comparing experience through visual behaviour in physical and virtual environments

David Greenwood and Oliver Jones

Experimental research in the built-environment disciplines is generally limited to the areas of materials and structures, and in the construction management field it remains relatively rare. In this example, an experimental study measured and compared subjects' experience and visual behaviour while viewing both physical and virtual environments. The epistemological stance is post-positivist, and the research approach is deductive. The research strategy was novel, combining the more common approaches to investigating experience (i.e. using self-reported semantic differential scales) with the physical capture and analysis of participants' visual behaviour using eye-tracking technology. In addition to conclusions on how far the virtual visual experience equates to that of a physical environment, this mixed approach to data collection enabled a comparison between the two sources: the one qualitative and the other quantitative. The first study involved using structured surveys of ninety respondents in order to develop semantic differential scales. These scales formed the basis of the second study, which was experimental and relied upon the two aforementioned data-collection strategies: (1) self-reporting by the participants using the semantic differential scales, and (2) the simultaneous capture and analysis of participants' visual behaviour, using a desktop eye-tracking tool. There was additional novelty in the analysis of the results and their presentation, which included box and whisker plots and 'heat map' visualisations to show eye-fixation frequency and viewing durations. The findings showed that virtual environments with higher levels of realism elicited experiential responses and visual behaviour that were closest to those observed with actual depictions of the physical environment. Although based within the discipline of architecture, the study drew upon theory and methods from fields such as cognitive psychology. It demonstrates how such methods can assist better understanding of human–environment interaction and examination of the validity of using virtual representations to recreate actual physical environments.

5.1 Introduction

Virtual environments of differing levels of realism are increasingly being used within the built environment for decision-making and communication, as they can offer early and inexpensive proxies of how a real environment will be experienced by its users (see, for example, Greenwood *et al.*, 2011). There have, however, been relatively few attempts to determine the extent to which the virtual visual experience equates to that of a physical environment. Almost all of these have been based upon self-reporting scales and rely on the subjects' recall of their experience rather than any more objective measures. This, in turn, raises questions as to whether virtual environments are, in the words of Rohrmann and Bishop (2002: 319), 'realistic enough to induce responses which are sufficiently similar to the evaluation of the real environment'. Methods that offer an opportunity for more substantive research do exist, though they originate in disciplines outside the built environment, and prospects for addressing these questions involve data-collection methods that have hitherto been more familiar in the fields of cognitive science, environmental psychology, physiology and artificial intelligence. An increased awareness of these methods would be beneficial for the built environment disciplines, in particular architecture, as information visualisation becomes both more advanced and more widely used, with a consequent need to understand matters such as how users process visual information in virtual environments.

In his 'Architecture of the senses', Howes (2005) recognises a number of leading studies in acoustics (Smith, 1999), olfaction (Corbin, 1986) and haptics (Harvey, 2003), but demonstrates that all of these support the dominance of the visual sense in the modern sensory model. Visual sensory information informs user perceptions, knowledge and experience formation, as well as influencing and reinforcing behaviours and actions. It may, therefore, be considered as key to informing critical decision-making. It is also argued (for example, by Ganah et al., 2001; Clayton et al., 2002; Bouchlaghem et al., 2005) that a greater understanding of visual behaviour and the way users gather and perceive visual information will lead to the development of improved virtual environments that can be used for better-informed design decisions, as well as improving the quality of design communication.

In their paper 'Perspective on computer aided design after four decades', Mark et al. (2007) identify eight 'approaches to design'. These approaches also represent eight possible avenues of computer-aided architectural design research. It is the last of their eight approaches, design and cognition, that is most relevant here. According to the authors, this approach 'bridges the disciplines of cognitive science, artificial intelligence and computer based methods. Research methods include capturing and observing human design activity and providing insights into human interaction with design tools or architectural places' (Mark et al., 2007: 6). The case study presented here is a description of such an attempt.

This study investigated the formation of experience during human–environment interaction with both physical and virtual environments. It aimed to address the following questions:

1   How is experience formed during human–environment interaction?
2   When physical environments are recreated using computer-generated virtual environments, do these evoke comparable visual experiences?
3   What effect do the different degrees of realism that are available in virtual environments have on that experience?

There was a mixture of methodological approaches and data-collection tools within the study. Importantly, this approach enabled a comparison to be made between evidence gathered from the more conventional self-reports of experience, obtained by questioning respondents after they had been confronted with controlled visual displays, and that gathered through the physical capture of subjects' visual behaviour using eye-tracking technology.

5.2 Research philosophy and research approach

Positioning this work in its ontological and, particularly, its epistemological context holds a certain irony, as the very subject under investigation is how knowledge and experience are constructed and understood in two separate conceptions of reality: that is, real and virtual environments. In broad terms, an ontological realist (accepting that there is a reality that is independent from our thinking about it) and epistemological empiricist stance was taken, together with a post-positivist approach to scientific enquiry: that is, one that strives to uncover objective 'truth', while simultaneously recognising the imperfection of pure theory. This adheres to what Ryan (2006: 18) describes as 'a learning rather than a testing role'. Post-positivism, as its name implies, starts from a positivist stance based on ontological realism and the ability to discover objective truth about this reality through deductive logic and the 'scientific method' of subjecting preconceived theory to empirical testing. The approach entails scepticism about the fallibility of perception (avoiding the extreme subjectivist stance that there *is no* external reality) and compensates for this fallibility by using multiple sources of evidence. Thus, post-positivism is often referred to as critical realism.

The following case study example is typical of the post-positivist philosophy in that it starts by borrowing from the constructivist tradition, developing an understanding of how people construct their perceptions of the world. In this case, the study set out to assemble sets of experiential and spatial descriptors that respondents used to articulate what they saw in various depictions of interior environments. It then proceeds in positivist mode by subjecting these constructions to empirical testing using experimental methodologies. The logical approach is, thus, first inductive and then deductive.

5.3 Data collection strategy and methodological choice

A multi-methods approach was adopted that combined both qualitative and quantitative data. In fact, there were three different phases to the study that represented three different methodological traditions. As there was no intention to reflect any sense of development over time, these were conducted over a cross-sectional, rather than longitudinal, time horizon. The study combined the following:

1   The qualitative social research methods of semi-structured face-to-face surveys (as exemplified in Bryman, 2012: 488) were employed to form a basis for the continuation of the work. This part of the study, as described below, involved

the collection of perceptual evidence that enabled the construction of bipolar self-reporting scales for the next phase.

2   *Quasi*-quantitative research methods (i.e. questionnaire-based self-reporting scales) that are prevalent in psychological experiments (as described, for example, by Coolican, 2014: 193) were used to collect data from participants on what they saw and felt about a set of images with which they were presented.

3   Finally, purely quantitative experimental data were collected using eye-tracking methods that are more common in physiological research (see, for example, Duchowski, 2002, 2007). This enabled the collection of objective data about the eye movements of participants as they viewed the different images of interior environments.

The use of 'mixed' or 'multi-' methods of data collection is not uncommon in built environment research, as the latter often involves cognitive, affective and behavioural components. This renders a single standard approach to data collection ineffective, as recognised by Amaratunga *et al.* (2002), who propose adopting a mixed methods approach in order to explore all of these components properly. McGrath (1982) has referred to the use of mixed methods as a 'necessary com-promise', whereas Yin (1994) regards it in a more positive light, counterbalancing the strengths and weaknesses of traditional polarised approaches and contributing to a form of methodological 'triangulation'.

In this study, a multi-methods approach was adopted, not only as an attempt to make the findings more robust, through triangulation involving data gathered from participants' self-reports (item 2, above) as well as objective measurement of their visual behaviour (item 3, above), but also to allow comparison (in the case of 2 and 3, above) between data collected through participants' reported experience and observed physical phenomena, as permitted by eye-tracking apparatus.

The study began with an open-ended survey (of ninety participants) to collect experiential and spatial descriptors that were stimulated by viewing of twenty images of varying interior environments. We shall refer to this as Phase 1 of the study, and more detail of the way it was administered will be given later. The descriptors were then thematically analysed and reduced to produce semantic differential scales with appropriate opposing terms: a method that, following its introduction by Osgood (1957), has been commonly used in the fields of psychology and physiology to capture data that report users' perceptions. These scales were designed for use in the second, experimental part of the study to prompt self-reports of experience during exposure to visual environments of varying realism.

The experimental part of the study (Phase 2) involved a controlled visual trial where twelve participants were presented with environmental representations of varying realism. There were two components to this, and the participants were asked to report their emotional responses and visual perceptions using the semantic differential scales. This will be referred to as Phase 2a of the study, and more exact details will follow. At the same time, in what will be referred to as Phase 2b, eye-tracking equipment was used to capture the participants' visual behaviour. These *quantitative physiological* measures (i.e. Phase 2b) were intended to complement, and offer comparison with the more commonly used *qualitative psychological* measures (i.e. Phase 2a) that capture reported experience in the environment. Together, these psychological and physiological measures provided data that enabled the comparison of visual behaviour between physical and virtual environments during the perceptual process and the formation of the experience.

In simple terms, the aim of the study was to compare perceptions of real and virtual environments. In line with its methodological position, the study as a whole used a progressive multi-methods approach to data collection, with the intention of increasing 'the likelihood that the sum of the data collected will be richer, more meaningful, and ultimately more useful' (Preskill, cited in Johnson *et al.*, 2007: 121). As noted by Saunders (2012), participant selection and the extent of collectable data are dependent upon available resources. Such limitations normally require the identification of an accessible population from which to draw a representative sample. This, in turn, creates problems of 'generalisability' from the sample (via the accessible to the theoretical populations) and imposes a rationale on the process of participant selection.

The participants in Phase 1 of the study were ninety second-year students from a single UK university, aged 21–22, with a roughly equal gender balance of forty-nine males (54.4 per cent) and forty-one females (45.6 per cent) and a predominance (88 per cent) of 'White British' ethnicity. The experiments were run on campus, over four separate sessions. Participation was voluntary, and students received no extra credit for participating. The normal ethical procedures were adhered to throughout. Following this, Phase 2 of the study used a smaller, randomly selected sample of the Phase 1 participants, but excluded any subjects who had previous experience of working in virtual environments (including use of the software used to create the virtual environments in the study). Phase 2 utilised a 'repeated-measures' design in which the twelve selected participants were given a controlled sequence of four different exposures for each of the twenty-four scales, each of which was an environmental representation of a different degree of 'realism'.

Fuller details of this will be provided in due course. Participants' reactions to these exposures were recorded in two ways, and concurrently: first, in Phase 2a, by their reporting their emotional responses and visual perceptions using the semantic differential scales; and then, in Phase 2b, through the use of eye-tracking equipment to capture the participants' visual behaviour. Results from the semantic differential scales (Phase 2a) enabled an examination of experiential responses to the computer-generated virtual environments of varying degrees of visual realism. A richer analysis was made possible through a combination of these data with the physiological measurements from the eye-tracking equipment (in Phase 2b), and this also afforded the possibility of a comparison between the two types of evidence.

5.4 Data collection

Phase 1 of the study used a semi-structured survey to collect a large qualitative data set of spatial and experiential descriptors of visual environments. The initial requirement was to select the environment itself. For this, ten architects were asked for examples of a diverse range of museum and gallery spaces (selected because their *raison d'être* is visual), which provided the images used for the study. Ninety participants were asked to view and describe twenty images of different interior environments. This produced 4,217 individual spatial descriptors and 1,920 experiential descriptors. Data collection was by semi-structured, face-to-face survey with open-ended questions, and the equipment used was a digital projector, a laptop and writing equipment. The survey structure was simple, restricting the number of open-ended responses to each question and directing the participant to respond to each environment with between one and five spatial descriptors and one and five experiential descriptors of their choice.

Following the approach of Braun and Clarke (2006), the process of thematic analysis included five steps, namely: familiarisation, generating initial codes, identifying themes, reviewing themes, and defining and naming themes. A total of 4,217 spatial descriptors were then organised into forty sub-themes, which were themselves grouped thematically so that eleven overarching themes from the data set could be identified. These were: (1) form, (2) material properties, (3) spatial qualities, (4) aesthetic judgement, (5) light, (6) order, (7) wayfinding, (8) function, (9) sensory interaction, (10) human characteristics, and (11) prospect and refuge. A similar process was undertaken with the 1,920 experiential descriptors, reducing them to fifteen themes: (1) relaxation, (2) attentiveness, (3) happiness, (4) context relevance, (5) fear, (6) boredom, (7) stress, (8) hostility, (9) sadness, (10) confidence, (11) preference, (12) empathy, (13) surprise, (14) shyness and (15) guilt.

Following the thematic analysis of Phase 1 data, a 'general classification model of emotion during human–environment interaction' was produced and subsequently refined. For further details of this model, its creation and its refinement, refer to Jones (2016). However, of more relevance to the methodological aspects of the process was the use of these data in the formation of a research instrument for Phase 2a of the study. This involved the extraction and use of 'descriptor pairings' to form twenty-five (twelve spatial and thirteen experiential) seven-point, bipolar, semantic differential scales. Further details of the process of transformation from more than 6,000 unique descriptors to these twenty-five semantic differential scales can be found in Jones (2016).

Phase 2a of the study used these scales to collect data on respondents' ($n = 12$) perceptions of real and virtual environments, and Phase 2b concurrently measured their visual behaviour using eye-tracking. The latter enabled the collection of objective quantitative data relating to physiological responses to visual stimuli, classified by descriptors such as 'fixation frequency' and 'fixation duration'. The Phase 2b eye-tracking data-collection and -analysis procedures followed the approach used by Blascheck *et al.* (2014).

Images of the selected test environment (a gallery space at the university) were reproduced in four stages of increasing realism. Three of the stages were recreated using computer-generated conditions of assumed increasing realism (namely, and in increasing order of realism, Google SketchUp, Autodesk Revit Architecture and Crytek CryEngine3), and the fourth condition (representing the '*real*' physical environment) was an exact photographic portrayal. All of the environments were depicted from exactly the same viewpoint and with the same field of view, using a process known as 'camera matching' (Prosser *et al.*, 2008). In addition to the semantic differential scales, described earlier, the equipment used comprised a powerful laptop computer with 18-inch monitor and associated Tobii T60 XL eye tracker. Tobii Analytics SDK (Analytics Software Development Kit) and Microsoft Excel were used in the subsequent analysis.

As described earlier, the twelve participants in Phase 2 were selected from the ninety who took part in Phase 1 of the study. Each participant was tested individually in a small office on campus. They were seated at a desk in front of the laptop equipped with the eye tracker. Each was given instructions that explained the experiment and that they were about to view twenty-five images under one condition. The same image was displayed twenty-five times, each time with one of the semantic differential scales attached. After viewing each image for 15 seconds, the participants completed the associated scale. The experiments were run over four separate sessions, once for each of the four conditions of assumed increasing realism (see above). Measures were taken to avert or mitigate the potential confounding variables that had been identified in the literature on visual experimentation. These included *viewpoint height and field* (Franz, 2005), *exposure duration* (Keul and Kühberger, 1997) and *task prescription* (Yarbus, 1967).

Results from the semantic differential scales during this study enabled the comparison of experiential responses to computer-generated virtual environments of varying degrees of visual realism. When combined with the physiological measurements from the eye-tracking data, this enabled a more detailed comparison of visual behaviour and experience during the consideration and completion of each semantic differential scale across each of the environments. Responses to the computer-generated conditions were then compared with the depiction of the physical environment.

5.5 Data analysis

There were two parts to Phase 2 of the study, the first employing respondents' perceptual self-reports using semantic differential scales, and the second employing physical measurements using eye-tracking equipment.

Phase 2a study: perceptual self-reports

Thematic analysis was used in Phase 1 of the study to create semantic differential scales from a large, bespoke set of spatial and experiential descriptors. This has already been described in Section 5.4. The present section will concentrate on the analysis of data from Phase 2 of the study. The *twelve spatial* and *thirteen experiential* semantic differential scales are here collectively referred to as 'visual experience' (VEx) scales.

For Phase 2a, each participant completed the VEx scales for each of the twenty-five images under each viewing condition. This procedure was repeated on a separate occasion for each of the four viewing conditions (each viewing condition signifying a different level of realism). Sets of results that combined each of the four viewing conditions were assembled: two examples are shown in Figure 5.1; these show the 'descriptor pairings' and map results for all four viewing conditions (i.e. SketchUp, Revit, Cryengine3 and Photo).

The numerical values recorded by each participant, for each condition and each scale, were input into Microsoft Excel using a QI Macros Excel 'plug-in'. *Box and whisker plots* were produced for each VEx scale to indicate the differences between the four viewing conditions.

The plots showed the range of responses, upper and lower quartiles, outliers and median lines, allowing comparison of responses for each condition. Charts were then produced that summarised the above results and enabled the overall comparison of each computer-generated condition with Condition 4 (the photographic portrayal). Figure 5.2 shows, for the experiential response data, the collected instances where each particular condition was most/least similar to Condition 4.

The chart shows that computer-generated Condition 3 (CryEngine) was consistently the most similar condition, followed by Conditions 2 (Revit) and 1 (SketchUp). This examination of experiential responses was then repeated using the *spatial* response data. Analysis of the spatial responses revealed comparable similarity patterns, with Condition 3 (CryEngine) showing 'most similar' to the photographic representation, at 60 per cent, followed by Conditions 2 (Revit) and 1 (SketchUp).

Phase 2b study: eye-tracking measurement

In Phase 2b of the study, eye-tracking data were captured using a Tobii desktop eye tracker and imported into the proprietary Tobii Analytics SDK software, where they were processed using automated aggregation analysis within the software to produce 'heat map' visualisations illustrating *fixation frequency* and *fixation duration* for each scale and each condition. This process resulted in 200 such heat map visualisations. An example is shown in Figure 5.3.

The process of constructing and modifying heat maps from eye-tracking data is discussed by Špakov and Miniotas (2007). For purposes of analysis, the scene used in Phase 2b was organised into regions denoted by the physical elements within the scene (considered to be more practicable than simply superimposing a grid), as shown in Figure 5.4.

From the results of the eye tracking, each of the nineteen regions in each of the 200 heat map visualisations was assigned a numerical score: that is 0 (no activity), 1 (low fixation frequency/duration), 2 (medium fixation frequency/duration) and 3 (high fixation frequency/duration). The numerical scores for each region were then input to Microsoft Excel to produce fixation frequency/duration trendlines (6th-order polynomial) for each scale under each condition. These were particularly useful for visualising patterns of visual behaviour and observing and comparing trends within the data.

From Phase 2b data, a chart was produced (similar to that shown in Figure 5.2) that enabled the comparison of the number of instances where each computer generation showed most similarity to Condition 4 (the photographic portrayal) in terms of fixation frequency and fixation duration. This is shown in Figure 5.5.

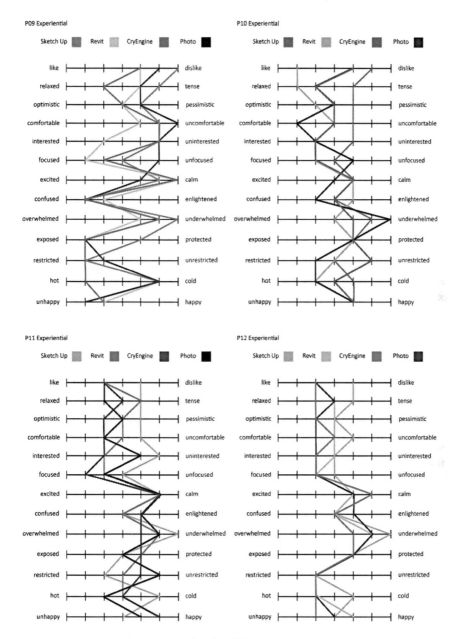

Figure 5.1 Two examples of VEx scales after all four viewing conditions

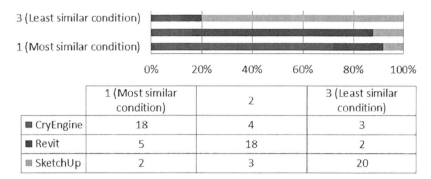

	1 (Most similar condition)	2	3 (Least similar condition)
■ CryEngine	18	4	3
■ Revit	5	18	2
▨ SketchUp	2	3	20

Figure 5.2 Similarity of computer generation to the depiction of the physical environment (experiential)

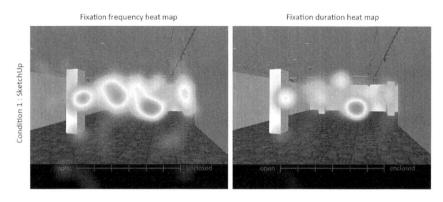

Figure 5.3 Example of heat map visualisation, showing fixation frequency and duration for a single viewing condition (SketchUp)

This followed the same pattern as the charts that compared similarity between the semantic differential scales (shown in Figure 5.2), namely, that computer-generated Condition 3 (CryEngine) was consistently the most similar condition, followed by Conditions 2 (Revit) and 1 (SketchUp).

5.6 Study findings and conclusions

The main findings of this research will be discussed here from three aspects, namely:

1   conclusions drawn from the participants' self-reports of visual perceptions using the semantic differential scales (Phase 2a);
2   outcomes from the analysis of eye-tracking data (Phase 2b); and
3   reflections on the comparison between evidence from the self-reports and that obtained from the eye tracking.

The findings from the self-reported (VEx) semantic differential scales suggest that the initial assumptions of increasing realism about the three computer-generated virtual environments were correct. The aggregated measures of both the experiential and spatial perceptions of participants confirmed this. In both cases, responses to the virtual environment created in CryEngine were most frequently closest to that of Condition 4 (the photographic portrayal). It was possible to calculate that the computer-generated condition of the lowest assumed visual realism, Condition 1

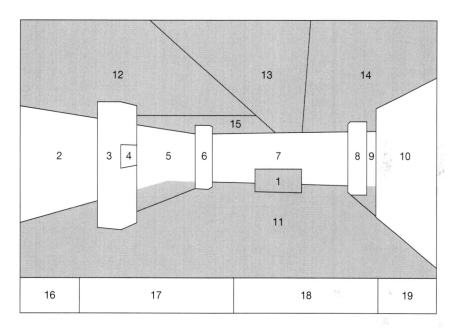

Figure 5.4 Nineteen regions identified from architectural elements within the scene

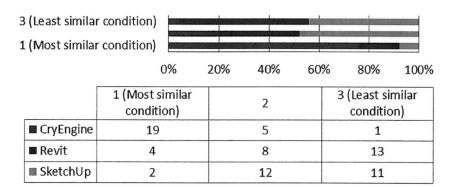

	1 (Most similar condition)	2	3 (Least similar condition)
■ CryEngine	19	5	1
■ Revit	4	8	13
▨ SketchUp	2	12	11

Figure 5.5 Similarity of computer generation to the depiction of the physical environment (aggregated fixation duration intensity)

(SketchUp) had, at best, only 58 per cent similarity to the depiction of the physical environment (Condition 4), whereas the computer-generated conditions of higher visual realism (Conditions 2 Revit and 3 CryEngine) achieved up to 83 per cent similarity.

5.7 Summary and conclusions

An analysis of the visual behaviour in terms of fixation frequency and fixation duration, carried out in Phase 2b of the study, confirmed the assumptions of increasing realism (as illustrated in Figure 5.5), but the physical capture of subjects' visual behaviour using eye-tracking technology also permitted a more objective comparison to be made, using the device illustrated in Figure 5.4, which 'zoned' the scene used in the study. Patterns of *fixation frequency* across all conditions were similar to patterns of fixation frequency identified in the depiction of the physical environment (Condition 4), though computer-generated virtual environments of higher visual realism evoke patterns of fixation frequency of greater similarity. In terms of the intensity of *fixation duration*, all computer-generated conditions exhibit distinctly similar traits to the depiction of the physical environment (Condition 4) with regard to visual behaviour across each region in the scene. Additional observations show the trendlines for the computer-generated conditions of higher visual realism were the most similar to the depiction of the physical environment.

In summary, the study produced a tentative, rather than definitive, answer to the question of whether computer-generated virtual environments are realistic. It was found that they do, to an extent, induce responses that approximate to those experienced in real environments, although, rather as expected, the similarity of response increases with higher visual realism. The area requires much more investigation. Arguably, the most significant contribution of the current study was a methodological one. Gaining a good understanding of how humans experience and respond to their environments is a key issue for the designers of buildings and infrastructure. At the same time, the interplay between what we see, what we think, and how we process and report these observations is complex. Conventional reliance on users' accounts of their experience may not be entirely sufficient. The research described here has shown how a diverse and unusual combination of research approaches and data-collection tools, drawing from a number of different research traditions, may lead to better understanding. It is hoped that this example will clarify the advantages of multi-method research approaches and encourage their adoption by future researchers, including postgraduate research students.

References

Amaratunga, D., Baldry, D., Sarshar, M. and Newton, R. (2002). 'Quantitative and qualitative research in the built environment: Application of "mixed" research approach', *Work Study*, 51(1): 17–31.

Blascheck, T., Kurzhals, K., Raschke, M., Burch, M., Weiskopf, D. and Ertl, T. (2014). 'State-of-the-art of visualization for eye tracking data', Proceedings of EuroVis, Swansea, UK, June 2014.

Bouchlaghem, D., Shang, H., Whyte, J. and Ganah, A. (2005). 'Visualisation in architecture, engineering and construction (AEC)', *Automation in Construction*, 14(3): 287–95.

Braun, V. and Clarke, V. (2006). 'Using thematic analysis in psychology', *Qualitative Research in Psychology*, 3(2): 77–101.

Bryman, A. (2012). *Social Research Methods* (4th edn). Oxford, UK: Oxford University Press.

Clayton, M., Warden, R. and Parker, T. W. (2002). 'Virtual construction of architecture using 3D CAD and simulation', *Automation in Construction*, 11(2): 227–35.

Coolican, H. (2014). *Research Methods and Statistics in Psychology* (6th edn). Hove, UK: Psychology Press.

Corbin, A. (1986). *The Foul and the Fragrant: Odor and the French social imagination*. Cambridge, MA: Harvard University Press.

Duchowski, A. (2002). 'A breadth-first survey of eye-tracking applications', *Behavioral Research Methods Instruments & Computers*, 34(4): 455–70.

Duchowski, A. (2007). *Eye Tracking Methodology: Theory and practice* (Vol. 373). London: Springer Science and Business Media.

Franz, G. (2005). 'An empirical approach to the experience of architectural space'. Unpublished PhD thesis, Max Planck Institute for Biological Cybernetics, Tubingen and Bauhaus University, Weimar.

Ganah, A., Anumba, C. and Bouchlaghem, N. M. (2001). 'Computer visualisation as a communication tool in the construction industry', *Fifth International Conference on Information Visualisation* (pp. 679–83). Washington, DC: IEEE Computer Society.

Greenwood, D. J., Lockley, S., Jones, O. G. F. and Jones, P. (2011). The efficacy of realistic virtual environments in capturing user experience of buildings. In P. Morand and A. Zarli (eds), *Proceedings of the CIB W78-W102 2011: International Conference*, Sophia Antipolis, France, 26–28 October.

Harvey, E. D. (ed.) (2003). *Sensible Flesh: On touch in early modern culture*. Philadelphia, PA: Pennsylvania Press.

Howes, D. (2005). Architecture of the senses. In M. Zardini (ed.), *Sense of the City: An alternate approach to urbanism* (pp. 322–31). Montreal and Baden: Canadian Centre for Architecture and Lars Müller.

Johnson, R. B., Onwuegbuzie, A. J. and Turner, L. A. (2007). 'Toward a definition of mixed methods research', *Journal of Mixed Methods Research*, 1(2): 112–33.

Jones, O. G. F. (2016). 'Human experience and visual behaviour in physical and virtual environments'. Unpublished PhD thesis, Northumbria University, UK.

Keul, A. and Kühberger, A. (1997). 'Tracking the Salzburg tourist', *Annals of Tourism Research*, 24 (4): 1008–12.

McGrath, J. E. (1982). The study of research choices and dilemmas. In J. E. McGrath and R. A. Kulka (eds), *Judgement Calls in Research* (pp. 69–102). Thousand Oaks, CA: Sage.

Mark, E., Gross, M. and Goldschmidt, G. (2007). 'A perspective on computer aided design after four decades', *Proceedings of eCAADe*, 26(04): 169–76.

Osgood, C. E. (1957). *The Measurement of Meaning* (No. 47). Urbana and Chicago, IL: University of Illinois Press.

Prosser, B., Gong, S. and Xiang, T. (2008). 'Multi-camera matching using bi-directional cumulative brightness transfer functions', *BMVC*, 8: 164–74.

Rohrmann, B. and Bishop, I. (2002). 'Subjective responses to computer simulations of urban environments', *Journal of Environmental Psychology*, 22(4): 319–31.

Ryan, A. B. (2006). Post-positivist approaches to research. In *Researching and Writing Your Thesis: A guide for postgraduate students* (pp. 12–26). Maynooth, Ireland: National University of Ireland Maynooth.

Saunders, M. (2012). Choosing research participants. In G. Symon and C. Cassell (eds), *Qualitative Organizational Research: Core methods and current challenges* (pp. 35–52). London: Sage.

Smith, B. R. (1999). *The Acoustic World of Early Modern England: Attending to the O-factor.* Chicago, IL: University of Chicago Press.

Špakov, O. and Miniotas, D. (2007). 'Visualization of eye gaze data using heat maps', *Electronics & Electrical Engineering*, 2: 55–8.

Yarbus, A. L. (1967). Eye movements during perception of complex objects. In *Eye Movements and Vision* (pp. 171–211). New York: Springer.

Yin, R. (1994). *Case Study Research: Design and methods.* Beverly Hills, CA: Sage.

6 Using quantitative approaches to enhance construction performance through data captured from mobile devices

Zeeshan Aziz, Christopher Barker and Bulent Algan Tezel

This chapter presents a quantitative analysis of production data captured through mobile devices on a wide range of projects, undertaken by a multinational infrastructure and services company. Lack of primary empirical data within the architecture, engineering and construction industry has, in the past, hindered developments of academic research in the field of performance measurement in construction. This chapter presents quantitative analysis of construction production data captured through mobile information systems and provides empirical evidence that data captured through mobile devices has the potential to deliver new and enhanced methods for performance measurement and enhancement. Relational data gathered through mobile devices are used to generate metrics against which construction issue resolution performance is measured. The chapter also discusses various methods for early identification and visualisation of performance deviations. The research approach and findings can be used for the development of academic performance measurement frameworks and also as an evidence base for further development by industry in the field of performance enhancement. The chapter contributes insight regarding innovative ways to interrogate construction production data and provide stimulus to others to develop the methods and approaches taken.

6.1 Introduction

Recent reviews of construction productivity performance indicate that the industry fell short in comparison with manufacturing and services-based industry sectors. Some of the key factors hampering construction productivity include issues with quality, use of project controls and adequate levels of supervision (Merrow *et al.*, 2009). Similar observations have been made in the UK Construction Industry Performance Report (Glenigan, 2014), indicating that the majority of construction projects continue to fail timely completion. This is coupled with falling profitability and client dissatisfaction with regards to product quality, service and value for money. Although quantitative data presented in performance reports are subject to interpretation, it is obvious that there is tremendous potential for productivity growth within the construction sector.

A review of academic literature and construction industry reports highlights inadequacies in terms of forecasting project costs, duration and other issues, and the inability of construction contractors to deliver quality products and services within a resource-constrained environment. A need for process innovation to help the industry deliver greater productivity and quality has long been identified. Also, the literature identifies a need for more accurate data relating to on-site construction activities, and to develop process intelligence through better use of project data, to help improve the accuracy of project planning. The literature provides a considerable amount of existing research identifying weaknesses in current performance measurement methods, including lack of availability of empirical data.

Recent developments in information communication technologies (ICT) make available a large and detailed digital sample of production performance data to test academic/theoretical hypotheses. After a review of academic research in the field of total quality management (TQM) and its evolution from Deming's management method, Rungtusanatham *et al.* (2003) make reference to 'the power of primary data'. One of the issues they identify is the restriction on further academic progress in the field of TQM without first validating Deming's principles against a sample of good empirical data. They suggest that, until they can move beyond secondary analysis of potentially inaccurate and 'weak' data, this will continue to stifle developments in this field.

With the development of information technologies (IT), the spectrum of automatic quantitative data-capturing systems has been widening. IT-based data-capture systems minimise human intervention, bias and error in the data-collection process. One of the important automated data-collection implications in construction is the use of sensors. Sensors convert a physical parameter (e.g. temperature, distance, humidity, displacement, flow, etc.) into an electronically measurable signal. Specific technologies such as radio frequency identification, near-field communication and Bluetooth beacons (nodes) are used for the automatic identification and monitoring of construction materials, equipment, plant and personnel. A global positioning system and a geographic information system can present data related to geography and location to the researcher. By providing an accurate point cloud at a reasonable speed, 3D laser scanning applications (i.e., the LiDAR and LADAR technologies) automatically capture data on the shape/features of a surface, space or topography. The technical capabilities, IT architecture and cost and the researcher's experience in using those data-collection systems should be well defined.

The literature review identified a range of issues relating to the quality and accuracy of industry performance data and, consequently, the capability of any existing performance measurement methods reliant upon it. It identified that data relating to production issues and performance are disparate, inconsistent, often subjective and highly retrospective. There is also evidence to suggest a lack of much-needed empirical primary data relating to construction performance. The main findings from the literature review are highlighted in Table 6.1.

Table 6.1 Performance measurement issues identified from literature review

Issue identified	Issue description	Supporting references
Consistency and standardisation	A lack of standard relational data for performance measurement and organisational learning Inability to identify relationships and patterns between performance outcomes and on-site decision-making	Dissanayake and Fayek (2008); Cheng and Wu (2012)
Complexity and data quality	The subjective nature of performance measurement data; not enough detail, visibility or science surrounding performance measures A lack of detail, accuracy and scale for performance metrics	Akhavian and Behzadan (2012); Cheng and Wu (2012)
Context and communication	Poor context for causes relating to project performance failures A lack of detailed industry metrics to measure performance against benchmarks	Son *et al.* (2012)
Lagging metrics and metrics	Retrospective measurement of performance outcomes that incur extensive lag time A lack of forecasting and lead performance measures	Barber (2004)
Visualisation	A need for improved analytical reasoning through visualisation, for project performance issues	Russell *et al.* (2009)

6.2 Research approach

Before the details of the research approach are explained, the particular research philosophy, which contains important assumptions about the researchers' view of the relationship between knowledge and the process by which it is developed, should be clarified. The adopted research philosophy will indicate four important underlying assumptions (Saunders *et al.*, 2009): (1) the ontological stance as the researchers' view of reality or nature of being; (2) the epistemological stance as the researchers' view on what constitutes acceptable knowledge; (3) the axiological stance as the researchers' view of the role of values in research; and (4) the data-collection techniques as the means to obtain data to generate knowledge.

The research philosophy adopted in this particular research is positivism. Positivists believe that reality is stable and can be observed and described from an objective viewpoint (Punch, 2005) – that is, without interfering with the phenomena being studied. Therefore, ontologically, reality is external, objective and independent of social actors. To positivists, reality must be investigated through the rigorous process of scientific enquiry. The positivist philosophy calls for focusing on facts and locating causality between variables (Easterby-Smith *et al.*, 2012). Thus, epistemologically, only observable phenomena can provide credible data. As reality is external and out of researchers' control, axiologically, research is undertaken in a value-free way, independent of researchers' values (objective stance). Data-collection and -analysis methods are generally quantitative, from large samples, with highly structured data-collection approaches.

Along with the philosophical stance of a research, the nature of data accessible to researchers is also a defining parameter in research approaches. Generally, when numerical or quantifiable data, or 'hard data', are more readily accessible, as in this particular research, the quantitative research methodology is employed (Neuman, 2007). Quantitative research involves 'explaining phenomena by collecting numerical data that are analysed using mathematically based methods (in particular statistics)' (Aliaga and Gunderson, 2005). Quantitative research tends to be explanatory and generally provides 'snapshots' or instantaneous results, used to address questions such as 'what', 'how much', 'how many' (Fellows and Liu, 2015). Quantitative research allows for a study with more breadth, involving a greater number of subjects and enhancing the generalisability of the results with the capabilities of replicability and comparison with similar studies (Kruger, 2003). However, it doesn't generally yield an in-depth analysis of the studied phenomenon, as the results are limited, providing numerical descriptions rather than detailed narratives, with less-elaborate accounts of human perception (Bryman, 2012). The quantitative research approach is well suited for the deductive reasoning in which hypotheses are tested with experiments, statistical methods, observations, etc., for generalisable confirmations or rejections (top-down approach), as opposed to the inductive reasoning in which observations lead to theories (bottom-up approach; Blaikie, 2009). Quantitative researchers design studies that enable the testing of hypotheses, which are tentative explanations that account for a set of facts open to further investigation. Both the quantitative and qualitative research methodologies and data-collection methods can be simultaneously employed in the same research to take advantage of their particular strengths (multi-method research). The study outlined in this chapter, however, adopts a mono-method approach that solely exploits the quantitative research methodology and data-collection methods, as they permit inference from, and gathering of, a greater number of quantitative data elements in a relatively short time and at a relatively low cost (Balnaves and Caputi, 2001).

The research in question aims to develop an understanding of how emerging structured and detailed quantitative data, created and made possible using a plethora of mobile ICT devices, could be applied to meet the performance measurement requirements of construction operations. Key objectives included:

- assessment of whether mobile data can provide the detail, accuracy and scale for performance measurements and metrics;
- assessment of whether mobile data could reduce the lag times associated with the identification of poor performance; and
- assessment of whether mobile data could provide improved visualisation of project performance.

Those points identified in the key objectives constitute the starting point or the hypothetical stance of the quantitative deductive approach with some explorative motives.

As the quantitative research methodology generally follows a linear research path (hypothesis–data collection–analysis), speaks the language of statistics, with variables and hypotheses, and emphasises precisely measuring quantitative or 'hard' data and testing hypotheses that are linked to general causal explanations, one needs to employ a relevant quantitative data-collection method or data-collection means through which quantifiable data are obtained for further analysis and manipulation (Gray, 2004). Observations, tests/experiments, surveys (questionnaires), and archive, document and secondary-data (i.e. databases, company records, etc.) studies are the frequently used quantitative data-collection methods (Creswell, 2013). In this particular case, existing (secondary) construction production data were statistically analysed to determine their effectiveness to enhance the existing performance measurement methods. Secondary data provide researchers with a relatively quick and cost-effective data source that may otherwise not be acquired first hand. They also enable both cross-sectional (one specific point in time) and longitudinal (extending over a period of time) analysis (Vartanian, 2010). However, because secondary data were collected by another entity at some point in the past for another purpose, before the research effort, secondary analysts have no opportunity to influence the initial data-collection method in terms of data quality (completeness and consistency in the data set), bias and compatibility with their research aims (Smith and Smith, 2008). This research demonstrates a longitudinal study of a *secondary-data* source, with its specific limitations (e.g. data completeness) underlined in the discussion section.

The secondary-data source, production-data sample, used in this study for quantitative analysis contained a large number of tables and data fields. A data sample is the representative subset of the whole population, in this case all the construction productivity data. The number of records relating to key data variables used for query and analysis as part of this research has been outlined in Table 6.2. A variable is a particular characteristic of the studied phenomenon that varies or has different values. Variables are used to statistically test hypotheses or assumptions. It is obvious from Table 6.2 that there is a deficit in the number of 'date due' and 'date closed' records when compared with the total number of production issues, immediately highlighting a shortfall in the number of resolution periods available to contribute towards averages.

After obtaining the raw data sample, which requires relatively less effort with readily available secondary-data sources, the data-analysis process necessitates management and preparation of the data sample in terms of data coding, data entry and data consistency for further descriptive and inferential statistical analysis. The data-preparation and -analysis procedures for the study were largely automated to minimise human bias, error and intervention. The data sample was collated into a structured cloud database with the aid of a proprietary application that utilises mobile touch-screen tablets as the primary means for data capture (Figure 6.1). The database also has a web interface that allows the mobile data to be updated, amended and reviewed via a desktop computer. An export of the cloud data has been provided as a Microsoft Access database file. The database file is intended to be used for grouping and running calculations against large numbers of records.

Table 6.2 Count of data variables used for query and analysis

Project variables	
Total number of projects	34
Total number of project types/construction sectors	10

Geographical variables	
Total number of regions	7

Company and user variables	
Total number of companies reporting issues	716
Total number of creator roles/professions	8
Total number of users creating data	259
Total number of trade classifications	46

Issue variables	
Number of general issue types	5
Total number of production issues	149,733
Total number of date-due records	136,628
Total number of date-closed records	140,571

The detailed findings were exported to Microsoft Excel for more detailed analysis and visualisation. The approach requires the data sample to be updated via Standard Query Language (SQL) query to generate elapsed periods between the creation and resolution of issues. The resolution periods were then used to determine performance averages, standard deviations and z-scores. Averages were established by grouping issue records according to the content of geographical, trade and role variables.

The production issues contained in the sample have been recorded across multiple projects and deliberately structured to hold a number of standard data variables. The variables deemed to be of interest for this research included project type, issue classification, location, trade, dates relating to issue identification, required resolution and actual resolution, and roles and professions of the individuals capturing the issues. These variables were used as the basis for SQL queries, designed to separate records into groups, for further analysis and performance calculation.

The approach to quantitative analysis included generating a query and running it against the date entries in the database to calculate and create resolution periods, against each production issue, according to issue type, project type, creator role and trade variables. Time was employed as a standard unit (days) and used as a metric for performance measurement. Once defined, the averages were compared against their hierarchical benchmarks to establish any variance with performance averages. Further, statistical analyses were conducted to determine standard deviations for resolution periods, and, from this, z-scores were used to identify proximity to average resolution period. A z-score or a standard score indicates how many an element is from the mean. It is a standardised value that lets researchers

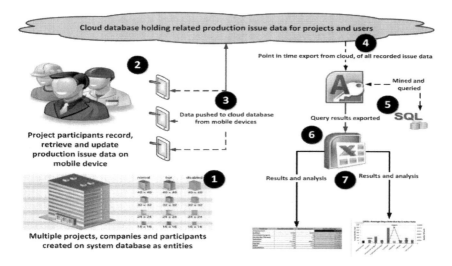

Figure 6.1 Data-collection and -analysis strategy

compare raw data values between different data sets, allowing the comparison of 'apples' and 'oranges' by the conversion of raw scores to standardised scores relative to population mean(s) (i.e. comparing one project's productivity values to another's productivity values; Field, 2009). This was identified as a proposed alternative and more robust statistical measurement for performance against a benchmark value. The following statistical analysis methods were used to undertake quantitative analysis.

Average resolution period

Determination of the average resolution period was the first step in quantitative analysis, to help evolve issue data into a standard measurement of performance. This was based on the premise that project performance is reliant on the timely resolution of production-related issues. Two averages were established for timely resolution of production issues:

1 *total resolution period*: the average time (in days) to resolve a production issue;
2 *overdue resolution period*: the average time (in days) in relation to the required resolution period.

Standard deviations for performance records

In this research, standard deviation values were used as an indication of the size of any variability, spread and distribution of resolution periods contained within the sample data. The larger the standard deviation, the more disparate and varied the sample data are assumed to be.

$$\sigma = \sqrt{\frac{\sum(X - \mu)^2}{n}}$$

$$Z = \frac{X - \mu}{s}$$

Standard Deviation
Formula

Z-Scores Formula

S	Total
s	Standard Deviation (Square Root of Variance)
μ	Average resolution period
n	Number of records
X	Individual Resolution period

Figure 6.2 Formulas used and legend to support analysis of average resolution, standard deviation and z-score calculations

Z-scores reflecting performance in relation to standard deviations

The z-score was used to determine how many standard deviations from the sample norm (average) any specific resolution periods were. If successful, this was used to evidence the capability to measure performance beyond simple mean calculations. It also confirms whether results sit above or below average resolution periods, owing to their positive or negative score. The z-scores were then converted to a percentage, to show where the records sit in respect to their performance against other data in the samples: that is, the percentage of records with quicker or slower response and resolution periods.

To structure and organise the analysis process better, the data sample was divided into smaller related data tables using SQL. SQL queries were named according to query type and its intended use and the level at which it groups any data.

Data distribution

The sample was queried to count the number of records by project type and region and to establish whether data were evenly distributed or clustered within geographical areas. The results indicate that records were clustered and not uniform, with heavier concentrations in the 'NTX' and 'WDC' regions, among projects types 'Education', 'Healthcare' and 'Other ancillary facilities'. The numerical distribution and 3D column chart are shown in Figure 6.3.

Project Type	Count of Issues	FLA	GEO	HOU	HSC	MHF	NTX	WDC
EDUCATION PREK-12	38580	14277	1405				22898	
HEALTHCARE	34782						8245	26537
HIGHER EDUCATION TRAINING	20983	20333					650	
OTHER ANCILLARY FACILITIES	20975			16				20959
MULTI FAMILY	15687				9538	6149		
OFFICE CORPORATE	10882	9608			974		12	288
AIRPORT	5983						5983	
PUBLIC USE CIVIC	1051	695			182		174	
POWER ENERGY	590							590
HOSPITALITY	220				220			

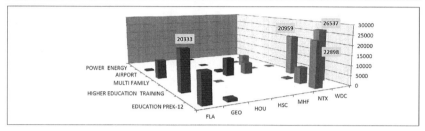

Figure 6.3 3D column chart showing distribution of issue records by project type and
region

Figure 6.4 identifies a dominant record set captured by the 'Contractor' having
provided 83.79 per cent of issues used to generate total resolution periods (117,787
of 140,571), and another majority related to general issues of the type 'Punch list',
with 85.8 per cent of issues (120,665 of 140,571).

Figure 6.5 represents the percentage of data entries that are missing for every
data field used to conduct SQL queries for analysis. This shows a generally good

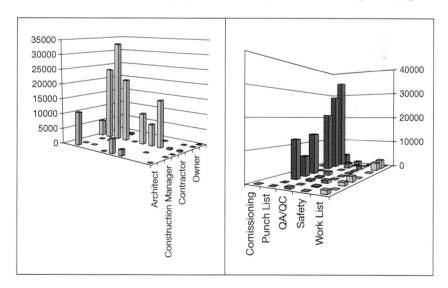

Figure 6.4 Record distribution by creator role and general issue type used to calculate total
resolution periods

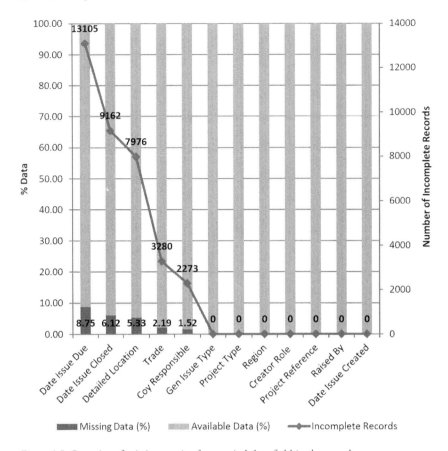

Figure 6.5 Quantity of missing entries for queried data field in the sample

level of information has been provided, with only 1.99 per cent of data fields (35,796 of a total 1,796,796) containing blank records. Most of the missing entries related to a date issue that was due or closed.

Z-score

A database query was created to establish the global organisation average and standard deviation for overdue resolution periods and total resolution periods from the data sample, to generate a metric for measurement that was statistically viable as an organisational performance standard. A further query of resolution periods for all issues was made against four projects. The data records were grouped by project, and the z-scores were then calculated for individual issue resolution periods, to establish the number of standard deviations from the data sample mean. The results were plotted into scatter graphs to help visualisation of general perform- ance distribution. The results of the distribution of the z-scores showed that two projects (i.e. PR20 and PR06) performed relatively well in the context of organisational delivery standards, with the majority of resolution periods sitting within ±1 standard deviation. The other two projects (PR15 and PR04) did not appear to perform very well, with large volumes of records creeping up to and beyond 2 standard deviations.

6.3 Discussion

Although the quantitative analysis approach relied on a data sample of considerable size, the content was inconsistent, and the levels of available information varied, depending on what part of the sample was being analysed. The data sample used was perfectly adequate for the research, with key fields related to who, when, where and what being available and available for analysis. However, when the research process was designed, it was initially presumed to establish time-based performance as the standard unit to measure performance. However, this approach was limited, owing to the number of missing date fields within the sample.

Quantitative analysis undertaken in this research has highlighted a number of development areas. It is apparent that the quality of the sample is largely dependent on the process used to prepare the system and collate the data. The sporadic and inconsistent volumes of data within the sample suggest that the organisation is using the platform in an ad hoc capacity, rather than as a standard, systematic means to capture and communicate production issues. The user base is also significantly biased towards contracting staff. The system used to capture the data may well benefit from more controls being employed in the data-capture process, to ensure fields such as date required and trade are made mandatory. Further standardisation of automated functions would also benefit the data sample. The removal of generic defaults for data entry such as 'other' would also help prevent large portions of intelligent data becoming unwieldy, as found with issue types. The number of unsolved items in the sample, together with the considerable lengths of time elapsed before the majority of issue are closed, again suggests process issues surrounding the implementation and management of the platform. It would appear that items are raised more quickly than they are closed, and that, perhaps, there is a lack of accountability for ensuring the system is updated.

This research set out to review the potential for mobile data capture to enhance performance measurement on construction projects. The research has identified potential areas and methods for further industry development where mobile data could be used to derive new methods for future performance benchmarking. There are new opportunities emerging from technology that allow organisations to gather and record production data quickly and on site. However, the research has shown that, although these technologies can provide a great deal of clarity and insight into the nature and cause of certain performance issues, there are still considerable limitations to a purely statistical approach. Also, results highlight that getting people to use the equipment correctly is the most important and pertinent issue at present. Although increased detail and improved methods of capturing detail surrounding performance issues will be of value to many, there is still the requirement for context and perception. This is not just an issue for construction: this is an issue for most industries, as technology pushes forward and allows us to measure things with greater degrees of accuracy.

References

Akhavian, R. and Behzadan, A. (2012). 'An integrated data collection and analysis framework for remote monitoring and planning of construction operations', *Advanced Engineering Informatics*, 26: 749–61.

Aliaga, M. and Gunderson, B. (2005). *Interactive Statistics* (3rd edn). Upper Saddle River, NJ: Pearson.

Balnaves, M. and Caputi, P. (2001). *Introduction to Quantitative Research Methods: An investigative approach*. London: Sage.

Barber, E. (2004). 'Benchmarking the management of projects: A review of current thinking', *International Journal of Project Management*, 22: 301–7.

Blaikie, N. (2009). *Designing Social Research* (2nd edn). Cambridge, UK: Polity.

Bryman, A. (2012). *Social Research Methods* (4th edn). Oxford, UK: Oxford University Press.

Cheng, M. Y. and Wu, Y. W. (2012). 'Improved construction subcontractor evaluation performance using ESIM', *Applied Artificial Intelligence*, 26: 261–73.

Creswell, J. W. (2013). *Research Design: Qualitative, quantitative, and mixed methods approaches* (4th edn). London: Sage.

Dissanayake, M. and Fayek, A. (2008). 'Soft computing approach to construction performance prediction and diagnosis', *Canadian Journal of Civil Engineering*, 35: 764–76.

Easterby-Smith, M., Thorpe, R. and Jackson, P. R. (2012). *Management Research* (4th edn). London: Sage.

Fellows, R. F. and Liu, A. M. M. (2015). *Research Methods for Construction* (4th edn). Oxford, UK: Wiley-Blackwell.

Field, A. (2009). *Discovering Statistics Using SPSS* (3rd edn). London: Sage.

Gray, D. E. (2004). *Doing Research in the Real World*. London: Sage.

Glenigan (2014). UK Construction Industry Performance Report 2014 [online]. Available at: www.glenigan.com/market-analysis/reports/construction-kpi-report-2014 (accessed 20 June 2015).

Kruger, D. J. (2003). 'Integrating quantitative and qualitative methods in community research', *The Community Psychologist*, 36(2): 18–19.

Merrow, E. W., Sonnhalter, K. A., Somanchi, R. and Griffith, A. F. (2009). Productivity in the UK Engineering Construction Industry. Reading, UK: Independent Project Analysis and Department for Business, Innovation & Skills.

Neuman, W. L. (2007). *The Basics of Social Research: Qualitative and quantitative approaches* (2nd edn). Boston, MA: Pearson Education.

Punch, K. F. (2005). *Introduction to Social Research: Quantitative and qualitative approaches* (2nd edn). London: Sage.

Rungtusanatham, M., Ogden, J. A. and Wu, B. (2003). 'Advancing theory development in total quality management: A "Deming management method" perspective', *International Journal of Operations & Production Management*, 23(8): 918–36.

Russell, A. D., Chiu, C.-Y. and Korde, T. (2009). 'Visual representation of construction management data', *Automation in Construction*, 18: 1045–62.

Saunders, M., Lewis, P. and Thornhill, A. (2009). *Research Methods for Business Students* (5th edn). Harlow, UK: Pearson Education.

Smith, E. and Smith Jr, J. (2008). *Using Secondary Data in Educational and Social Research*. Maidenhead, UK: McGraw-Hill Education.

Son, H., Park, Y., Kim, C. and Chou, J. S. (2012). 'Toward an understanding of construction professionals acceptance of mobile computing devices in South Korea: An extension of the technology acceptance model', *Automation in Construction*, 28: 82–90.

Vartanian, T. P. (2010). *Secondary Data Analysis*. New York: Oxford University Press.

Part III

Qualitative research

7 A theoretical framework for conserving cultural values of heritage buildings in Malaysia from the perspective of facilities management

Hasif Rafidee Bin Hasbollah and
David Baldry

This chapter presents a case study approach that gives in-depth coverage of the conservation process in Malaysia: the Acts, policy guidelines and documents used in the conservation process, and the facilities management practices used to conserve the cultural values of heritage buildings in Malaysia. The chapter illustrates a critical realist philosophy that aims to understand the world by distinguishing reality from factual and empirical knowledge. The study aimed to develop a framework for conserving the cultural values of heritage buildings from the perspective of facilities management in order to sustain the physical condition of a heritage building in Malaysia. The framework integrates a facilities management perspective (integration of people, place, process and technology) with heritage building conservation (mapping the eight criteria of cultural values of heritage buildings). The research identified the cultural values of heritage buildings as being social, economic, political, historical, aesthetic, scientific, age-related and ecological. These were associated and epistemologically constructed with the facilities management perspective of people, place, process and technology. Heritage buildings are conserved using a process called 'value-based management'. However, conflicts can occur as value clashes and goal incompatibility among the heritage stakeholders engaging in values-based management. The study proposed a systematic framework that helps to prevent the deterioration that leads to a magnitude of loss of heritage buildings in Malaysia. The development of the theoretical framework for conserving cultural values of heritage buildings from the perspective of facilities management in Malaysia was based on interviews at three levels of conservation practitioner (strategic, tactical and operational levels) and document reviews (National Heritage Act, Outstanding Universal Values, Conservation Guidelines and Conservation Management Plan). In this study, content analysis and template analysis were undertaken in order to obtain generalisations of the findings.

7.1 Introduction

A heritage building is defined as 'an historic building that gives us a sense of wonder and makes us appreciate culture and our heritage' (Feilden, 2003). This definition describes a heritage building as an object that is unique and distinctive, such that it is capable of attracting curiosity about its existence and the history behind its being. Thus, a heritage building is a part of human creation that produces an icon for a country, provides local identity, reflects cultural values and background, represents a source of memory of historical events and also contributes to the tourism business industry (UNESCO, 1972, 2003; Robinson, 2000; Feather, 2006; Smith, 2006; Timothy and Boyd, 2006; Timothy, 2007; ARCADE, 2008; Communities and Local Government, 2009; Woon and Mui, 2010; Loulanski and Loulanski, 2011).

In England, the process of conserving a heritage building begins with understanding and defining 'how', 'why' and to 'what extent' cultural values contribute to the local identity and resources for current and future generations (English Heritage, 2008). Cultural values are understood to be dynamic and perceived through different lenses, but what is critical is to accept the changeability and significant changes of values from one culture or period to another (Hall, 1997; Mason, 2006; Heras *et al.*, 2013). This could be seen when four distinctive cultural values of social, historic, aesthetic and scientific were established by UNESCO's World Heritage Committee (2008) and later were followed by economic, political, ecological and age-related, to complement the conservation development process (Piper, 1948; Reigl, 1982; Lowenthal, 1985; Riganti and Nijkamp, 2005).

Heritage buildings are conserved using a conservation process called 'value-based management' (VBM). ICOMOS (1999) has recognised VBM as the dynamic process by which heritage stakeholders conserve heritage sites and places (Clark, 1999, 2001; Kerr, 2000). VBM emphasises conserving and protecting the significance of the heritage site and place as defined by designation criteria, government authorities or other owners, conservation experts and other citizens with legitimate interest in the place (Mason *et al.*, 2003). There are four stages of the VBM process, which are to:

- understand the significance of cultural heritage;
- develop a policy of preserving cultural heritage;
- manage in accordance with policy; and
- change in accordance with policy.

The phrase 'heritage stakeholders' refers to individuals or groups who have a vested interest in heritage buildings. These normally consist of heritage-building owners, local communities, historians, conservation specialists and architects, heritage-building surveyors, governments and also non-governmental organisations (NGOs). However, conflicts occur as value clashes and goal incompatibility among the heritage stakeholders engaged in VBM emerge (Finlayson, 2011). Furthermore, conserving the cultural values of heritage buildings (CVHB) in VBM is not only potentially a prime context for conflict: that conflict also sits at the core of any attempts to deal with cultural heritage management (CHM) practice. CHM conflicts, such as discordance of interest among heritage stakeholders (for instance, government and NGOs), the domination of power (power to decide), political systems, ethnic and community disputes, and selective commodification, lead to loss of cultural heritage (Rowlands, 1994; Tunbridge and Ashworth, 1996; Meskell, 2002; Rowlands and Butler, 2007; Perring and Linde, 2009).

Therefore, the purpose of this chapter is to develop a framework for conserving CVHB from a facilities management (FM) perspective in order to sustain the physical condition of a heritage building. Thus, FM has been chosen because of its familiarity with the building care process. The framework was intended to integrate the FM perspective (integration of people, place, process and technology in the conservation of a heritage building) and heritage building conservation (including mapping the eight criteria of CVHB), so that the two are seen as one activity, rather than processes that occur at opposite ends of a spectrum. This systematic framework may help to prevent the deterioration that leads to a magnitude of loss of CVHB in Malaysia.

7.2 Research methodology

This research was influenced by the critical realist approach, which argues that 'real' social structures interact with individuals. Realism, in contrast to the propositions of empiricism and positivism, posits that the social world does not exist separately from humans and their interpretations of it, but is constructed by rules and procedures using their knowledge, understanding and connections (Bhaskar, 1989; Smith, 1998; Sayer, 2000). The critical realist explains social reality, criticises social order and understands people from the 'inside' in order to interpret the 'meaning' of human actions on social actions in the real world (Fay, 1980; Lamnek, 1988; Patton, 1990; Crabtree and Miller, 1992). The critical realist philosophy views understanding the world as distinguishing reality from factual and empirical knowledge and recognises the structures and mechanisms in the event or phenomenon. As a turning point, this research began to discover the literal meaning of the phenomenon by understanding human interaction and focusing on the conservation practitioners in Malaysia conserving CVHB from the perspective of FM.

The study adopts a case study approach, which is used in many situations to contribute to the knowledge of individuals, groups, organisations and social, political and related phenomena. It helps to retain the holistic and meaningful characteristics of real-life events such as organisational and managerial processes (Yin, 2003, 2009). This research lies in the realm of a single-case (embedded) design. The rationale for adopting a single-case design is that the case is an 'extreme' or 'unique' case, or it is a 'representative' or 'typical' case, and also the depth of coverage from a single case is adequate. This research explored the in-depth coverage of the conservation process in Malaysia: the Acts, policy, guidelines and documents used in the conservation process and the FM practices employed to conserve the

CVHB in Malaysia. The specific case that was studied was the seventeenth-century 'Stadhuys' or 'Red Building' in Malacca, Malaysia, which was built as the official residence of the Dutch governors and their officers.

7.3 Data-collection strategy

The development of the theoretical framework for conserving CVHB from the perspective of FM in Malaysia was based on interviews and document reviews. However, before the development framework of this research was mapped, it was vital to understand the data-collection strategy of the research. Figure 7.1 indicates the data-collection strategy.

Interviews

The interviews were used to explore the expert views and insights of the three levels of conservation practitioner:

1   *strategic*: the deputy commissioner of the Cultural Heritage Department and the director of Registration and Enforcement;
2   *tactical*: the director of the World Heritage Organisation of Malacca and the conservation architect from Malacca City Council;
3   *operational*: Curators/Conservators I and II of the Malacca Museum Corporation, the curator assistant and also the contractor/conservator appointed by the Cultural Heritage Department of the Heritage Department of Malaysia.

These eight respondents were chosen (two strategic respondents, two tactical respondents and four operational respondents) as they were purposively involved in the process of the conservation of CVHB in the state of Malacca, Malaysia (which was endorsed as having World Heritage Status by UNESCO).

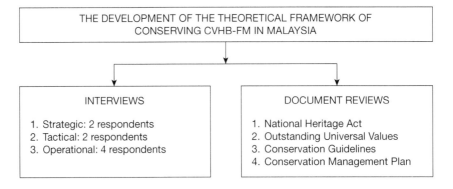

Figure 7.1 The data-collection strategy

Document reviews

The document reviews were used to analyse and review a variety of existing sources, with the intention of collecting independently verifiable data and information. Four documents were reviewed, as these documents were used in the process of conservation of CVHB in the state of Malacca. These documents were the National Heritage Act (Malaysia), UNESCO's Outstanding Universal Values (OUVs), Conservation Guidelines (Malaysia) and the Conservation Management Plan (CMP) for the state of Malacca, Malaysia.

7.4 Data-collection method

Polkinghorne (2005) stated that a qualitative method is most effectively used to investigate the human experience. Nevertheless, the quantitative method of conducting research suits a situation where the sample size is very large and can later be generalised to a large population, whereas qualitative research focuses on a particular subject in detail (Myers, 2009). This research did not involve a large sample size but was focused on several expert respondents who were purposively responsible for the conservation process in the state of Malacca. In this research, an interview technique was applied. The interview technique is commonly used in social science contexts and involves the researcher asking questions and receiving answers from the individual being interviewed (Robson, 2002; Sarantakos, 2005). Furthermore, the application of semi-structured interviews can cover a wide range of subject matter, using a series of questions that are in the general form of an interview schedule, but with the ability to diverge from the sequence of questions, should the opportunity arise (Bryman and Bell, 2007). Moreover, the semi-structured interview is considered to be advantageous because it can be modified based upon the interviewer's perception of what seems most appropriate (Sulaiman, 2012).

Interview guidelines were used to help the researcher direct the discussion towards the topics and issues relating to the conservation of CVHB from the perspective of FM in Malaysia. In this research, the interview guideline consisted of five main topics:

1   the respondent's professional background;
2   understanding of the current practice of conservation;
3   identifying the current FM perspective influencing the conservation process;
4   developing a theoretical framework for conserving CVHB from the FM perspective in Malaysia;
5   suggestions/opinions/views.

Data analysis

In any study or research, data analysis is critical and usually extensive. There is no standardised procedure for analysing data. According to Easterby-Smith *et al.*

(2002), to make the data collected meaningful for the study, a clear explanation of how the analysis is done and a demonstration of how the raw data are transformed into a meaningful conclusion are required. Burns (2000) asserted that the purpose of data analysis is to find meaning, and this is done by systematically arranging and presenting the information. As this research used an inductive approach, the verification of data generation, analysis and theory verification take place concurrently with the construction of a theory. In this research, content analysis and template analysis were undertaken in order to obtain generalisation of the findings.

Content analysis

The data gathered from the interviews and document reviews were analysed using content analysis. Content analysis is an important and powerful tool in the analysis of qualitative research and has a systematic technique to collate valid inferences from texts (or other meaningful matter) into fewer content categories, based on explicit rules of coding and themes (Stemler, 2001; Wilkinson and Birmingham, 2003; Krippendorff, 2004; Babbie, 2007; Saldana, 2009). For instance, in the content analysis, all the interview transcripts from the interviews were carefully considered to produce a sense of the whole data. The respondents' responses were extracted and brought together into one table, which constituted the unit of analysis. The table was divided into 'respondent identifier', 'interview text', 'interpretation of the underlying meaning' and 'descriptive codes'.

The context of the respondents' views and perceptions was complex and complicated to understand, and, therefore, the meaning of the 'interview texts' was condensed into an 'interpretation of the underlying meaning' that would be summarised and manifested in the content of the responses. The condensed 'underlying meaning of the interpretation' was seen as a whole and abstracted into 'descriptive codes'. The 'descriptive codes' were the thread of meaning running through the condensed text that was encrypted. In this research, the 'descriptive code' highlighted and represented the key features of the research. The 'descriptive code' was an epistemological criterion that manifested the variables of CVHB and FM perspectives that were used in generalising the objectives of the research.

Content analysis of the documents also consisted of four columns ('respondent identifier', 'interview text', 'interpretation of the underlying meaning' and 'descriptive codes'). Hence, the rationale for applying content analysis in this research was to provide insights directly, via texts and transcripts of documents taken verbatim from the respondents answering questions based on the objectives of the research. An example of content analysis of the expert interviews is shown in Table 7.1.

Template analysis

According to King (2006), template analysis is a method of thematically organising and analysing textual data that focuses on using the textual content to describe a

Table 7.1 Example of content analysis of interviews

Research objective 1: Understanding the current practice of conservation			
Respondent identifier	Interview text	Interpretation of the underlying meaning	Descriptive codes
Strategic: R_1	Preservation and conservation of heritage buildings in Malaysia are based on the National Heritage, the OUV, Guidelines for Conservation and also CMP. All the CVHB are under those documents	Conservation of heritage buildings in Malaysia is based on: • the National Heritage Act$_1$; • the OUV$_2$; • the Guidelines for Conservation of Heritage Building in Malaysia$_3$; and • the Guidelines for Conservation Management Plan of Malacca$_4$ All eight classifications of CVHB$_5$ are stated in these four documents	$_1$: NHA – D$_1$ $_2$: OUV – D$_2$ $_3$: GCHB – D$_3$ $_4$: CMP – D$_4$ $_5$: CVHB

phenomenon. Template analysis is used in the process of organising and analysing textual content (Crabtree and Miller, 1999). Furthermore, Saunders *et al.* (2009) assert that a list of categories, codes or templates represent the themes or issues revealed from the data that have been collected. These codes are very important in the interpretive process of developing a theoretical framework. In addition, template analysis is a flexible technique, with fewer specific procedures, that permits a researcher to tailor the procedures to match the requirements of their aim and objectives (King, 2006).

In this research, template analysis was adopted to develop the theoretical framework for conserving CVHB from FM perspectives. Therefore, three main phases of applying template analysis for this qualitative research were used. The process of applying template analysis began with the creation of an *initial* template.

Creating an initial template

Defining codes and clustering the themes were the two main processes in the development of the initial template analysis. According to King (2011), 'themes' are features of the respondent's perception or experiences relevant to the research question, and 'coding' refers to the process of identifying themes in accounts and attaching labels (codes) to index them. In this study, the 'themes' of this research were related to the research objectives, and the 'codes' were more specifically

FACILITIES MANAGEMENT PERSPECTIVES (FM)

People

- Leadership and Management Principles;

- Knowledgeable in Cultural Values of Heritage Buildings (CVHB); National Heritage Act Malaysia (NHA); Outstanding Universal Value (OUV); Guidelines for Conservation of Heritage Buildings (GCHB); and Conservation Management Plan (CMP).

Place

- Heritage buildings include the site, landscape and surroundings.

Process

- Conserving CVHB includes preservation, restoration, reconstruction, rehabilitation, and adaption or any combination method according to NHA, OUV, GCHB and CMP.

Technology

- Appropriate Technology as the mechanism and medium to assist the conservation activity.

Figure 7.2 Example of the *initial* template for FM perspectives

known as descriptive codes. Categorisation of descriptive codes depends on the volume of data, specificity or amount of detail needed for data analysis and generalisation of the study. Descriptive codes can be assigned more detailed coding as 'sub-codes', when needed (Miles and Huberman, 1994). Furthermore, Gibbs (2007) explained that the most general code is called the 'parent', and its sub-codes are the 'children'; hence, 'sub-codes' that share the same 'parent' are the 'siblings' in a hierarchy. Once again, only 'parent code' was applied in order to generalise the topics. For instance, the National Heritage Act is coded as NHA. Figure 7.2 illustrates an example of the *initial* template for FM perspectives for this study.

Revising the initial template

After the *initial* template has been completed, it can be developed until the researcher feels that it gives as good a representation as possible of the 'themes' identified in the data (King, 1998). According to King (2006), once the initial template has been constructed, the researcher needs to revise it in order to reveal any inadequacies that arise within the template. This will involve insertion, deletion, changing the scope and also adjusting the higher-order classification of a 'theme'. Revision of the initial template was completed in the validation phase of the research. The insertion, expansion, reduction and deletion of the 'code' occurred in this phase.

Developing a final template

According to King (2006), there is no stage at which the researcher can say with absolute certainty that the template is 'finished'. This is because there are always other ways of interpreting any set of qualitative data (King, 2011). Moreover, the decision on when to stop analysing data is a critical point among researchers (Abukhzam, 2011). Therefore, a 'considered' final template exists only when most or all transcribed data have been read through carefully and repeatedly, and when the researcher is confident that the template is accurate. Moreover, King (2006) asserts that the researcher may insert, expand, reduce and delete some of the material that was not successfully encompassed in the final template or change the template as and when necessary to know when to stop the development of the template. However, in this study, the final template was established after it had been validated and verified by the strategic-level respondents.

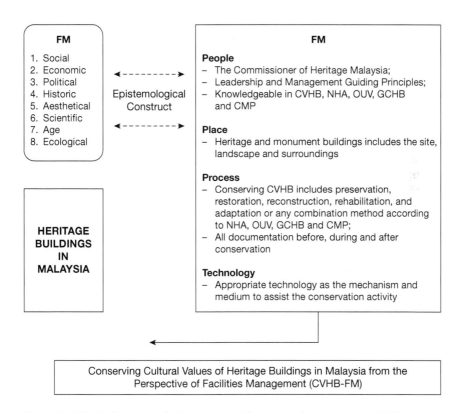

Figure 7.3 The final template for the theoretical framework for conserving CVHB in Malaysia from the perspective of FM

7.5 Main findings and conclusions

This research identified the CVHB as being social, economic, political, historical, aesthetic, scientific, age-related and ecological. These were associated and epistemologically constructed with the FM perspectives of people, place, process and technology. Figure 7.3 indicates the final template for the theoretical framework for conserving CVHB in Malaysia from the perspective of FM.

The embedded levels (strategic, tactical and operational) of the respondent conservation practitioners were explained and elaborated in connection with the characteristics of CVHB and FM in developing the theoretical framework for the research. The research provided insights into how the perspective of FM was associated with CVHB criteria in conserving a heritage building in Malaysia. It was hoped and intended that this theoretical framework would assist Malaysian practitioners in advancing the application of FM in the conservation of heritage buildings and their cultural values by highlighting the role of FM as a robust tool in conservation practice.

References

Abukhzam, M. F. (2011). *The Development of a Framework to Aid the Identification of Factors Inhibiting Bank Staff's Attitude Towards E-Banking Adoption in Libya*. PhD thesis, University of Salford, UK.

ARCADE. (2008). Preservation of Cultural Heritage and Local Community Development: History, identity and memory. Report from ARCADE (Awareness Raising on Cultural and Development in Europe), Stalowa Wola-Krakow, Poland, 29–30 May 2008. Available at www.ifacca.org/events/2008/05/28/preservation-cultural-heritage-and-local-community/ (accessed 6 December 2015).

Babbie, E. R. (2007). *The Basics of Social Research*. Belmont, CA: Wadsworth.

Bhaskar, R. (1989). *Reclaiming Reality*. London: Verso.

Bryman, A. and Bell, E. (2007). *Business Research Methods* (2nd edn). Oxford, UK: Oxford University Press.

Burns, R. B. (2000). *Introduction to Research Methods* (4th edn). London: Sage.

Clark, K. (1999). *Conservation Plans in Action*. London: English Heritage.

Clark, K. (2001). *Informed Conservation*. London: English Heritage.

Communities and Local Government. (2009). *Draft Planning Policy Statement 15: Planning for the Historic Environment*. Communities and Local Government Publications, London [online]. Available at: http://webarchive.nationalarchives.gov.uk/20120919132719/www.communities.gov.uk/documents/planningandbuilding/pdf/consultationhistoricpps.pdf (accessed 6 December 2015).

Crabtree, B. F. and Miller, W. L. (eds) (1992). *Doing Qualitative Research*. Newbury Park, CA: Sage.

Crabtree, B. F. and Miller, W. L. (1999). A template approach to text analysis: Developing and using codebooks. In B. F. Crabtree and W. L. Miller (eds), *Doing Qualitative Research* (pp. 163–77). Newbury Park, CA: Sage.

Easterby-Smith, M., Thorpe, R. and Jackson, P. R. (2002). *Management Research: An introduction* (3rd edn). Los Angeles, CA: Sage.

English Heritage. (2008). *Conservation Principles: Policies and guidelines*. London: English Heritage.

Fay, B. (1980). *Social Theory and Political Praxis*. London: Allen & Unwin.

Feather, J. (2006). Managing the documentary heritage: Issues from the present and future. In G. E. Gorman and J. S. Sydney (eds), *Preservation Management for Libraries, Archives and Museums* (pp. 1–18). London: Facet.

Feilden, B. (2003). *Conservation of Historic Buildings* (3rd edn). Oxford, UK: Architectural Press.

Finlayson, P. (2011). What is meant by Values-based Management? Housewright Building and Restoration, 31 October [online]. Available at http://oldhousewright.wordpress.com/ (accessed on 17 April 2012).

Gibbs, G. (2007). *Analysing Qualitative Data*. Book 6 of Sage Qualitative Research Kit. London: Sage.

Hall, S. (1997). *Representation: Cultural representations and signifying practices*. London: Sage/Open University.

Heras, V. C., Wijffels, A., Cardoso, F., Vandesande, A., Santana, M., Orshen, J. V., Steenberghen, T. and Balen, K. (2013). 'A value-based monitoring system to support heritage conservation planning', *Journal of Cultural Heritage Management & Sustainable Development*, 3(2): 130–47.

ICOMOS. (1999). The Australia ICOMOS Charter for the Conservation of Places of Cultural Significance [online]. Available at http://australia.icomos.org/wp-content/uploads/BURRA_CHARTER.pdf (accessed 9 January 20120).

Kerr, J. S. (2000). *Conservation Plan* (3rd edn). Sydney, NSW: National Trust of Australia.

King, N. (1998). Template analysis. In G. Symon and C. Cassell (eds), *Qualitative Methods and Analysis in Organisational Research* (pp. 118–34). London: Sage.

King, N. (2006). 'Professional identities and innovation: A phenomenological constructivist approach', *British Academy of Management Conference*, Belfast, Ireland, 12–14 September 2006.

King, N. (2011). Template analysis: Example 1: Carers' experiences of out-of-hours palliative care services [online]. Available at www.hud.ac.uk/hhs/research/template-analysis/example-1/ (accessed 6 December 2015).

Krippendorff, K. (2004). *Content Analysis: An introduction to its methodology* (2nd edn). Thousand Oaks, CA: Sage.

Lamnek, S. (1988). *Qualitative Sozialforschung*. Band 1: *Methodologie*; Band 2: *Methoden und Techniken*. Munich: Psychologie Verlags Union.

Loulanski, T. and Loulanski, V. (2011). 'The sustainable integration of cultural heritage and tourism: A meta-study', *Journal of Sustainable Tourism*, 19(7): 837–62.

Lowenthal, D. (1985). *The Past is a Foreign Country*. Cambridge, UK: Cambridge University Press.

Mason, R. (2006). 'Theoretical and practical arguments for values-centered preservation', *Cultural Resource Management: The Journal of Heritage Stewardship*, 3(2): 21–48.

Mason, R., MacLean, M. G. H. and Torre, M. D. L. (2003). *Hadrian's Wall World Heritage Site: English Heritage, a case study*. Los Angles, CA: The Getty Conservation Institute.

Meskell, L. (2002). 'Negative heritage and past mastering in archaeology' *Anthropological Quarterly*, 75(3): 557–74.

Miles, M. B. and Huberman, A. M. (1994). *Qualitative Data Analysis*. London: Sage.

Myers, M. D. (2009). *Qualitative Research in Business and Management*. London: Sage.

Patton, M. Q. (1990). *Qualitative Evaluation and Research Methods* (2nd edn). Newbury Park, CA: Sage.

Perring, D. and Linde, S. V. D. (2009). 'The politics and practice of archaeology in conflict', *Conservation & Management of Archaeology Sites*, 11(3–4): 197–213.

Piper, J. (1948). Pleasing decay. In *Buildings and Prospects* (pp. 89–116). London: The Architectural Press.

Polkinghorne, D. E. (2005). 'Language and meaning: Data collection in qualitative research', *Journal of Counselling Psychology*, 52(2): 137–45.

Reigl, A. (1982). 'The modern cult of monuments: Its character and its origin [1903]', *Oppositions*, 25(Fall): 21–51.

Riganti, P. and Nijkamp, P. (2005). 'Benefit transfers of cultural heritage values: How far can we go?', *Proceedings of the European Regional Science Association (ERSA) Conference*, European Regional Science Association, Amsterdam.

Robinson, M. (ed.) (2000). *Tourism and Heritage Relationships: Global, national and local perspective*. Sunderland, UK: Business Education.

Robson, C. (2002). *Real World Research: A resource for social scientists and practitioner–researchers*. Oxford, UK: Blackwell.

Rowlands, M. (1994). The politics of identity in archaeology. In G. C. Bond and A. Gilliam (eds), *Social Construction of the Past: Representation as power* (pp. 129–43). London: Routledge.

Rowlands, M. and Butler, B. (2007). 'Conflict and heritage care', *Anthropology Today*, 23(1): 1–2.

Saldana, J. (2009). *The Coding Manual for Qualitative Researchers*. London: Sage.

Sarantakos, S. (2005). *Social Research* (3rd edn). New York: Palgrave.

Saunders, M., Lewis, P. and Thornhill, A. (2009). *Research Methods for Business Student* (5th edn). Harlow, UK: Pearson Education.

Sayer, A. (2000). *Realism and Social Science*. London: Sage.

Smith, L. (ed.) (2006). *Cultural Heritage: Critical concepts in media and cultural studies*. London: Routledge.

Smith, M. (1998). *Social Science in Question*. London: Sage.

Stemler, S. (2001). 'An overview of content analysis', *Practical Assessment, Research & Evaluation*, 7(17) [online]. Available at http://pareonline.net/getvn.asp?v=7&n=17 (accessed 5 April 2012).

Sulaiman, N. (2012). *Opportunities for the Transfer of United Kingdom Best Practices for the Provision of Public Residential Care Facilities for the Elderly to Malaysia*. PhD thesis, University of Salford, UK.

Timothy, D. J. (ed.) (2007). *Managing Heritage and Cultural Tourism Resources*. Aldershot, UK: Ashgate.

Timothy, D. J. and Boyd, S. W. (2006). 'Heritage tourism in the 21st century: Valued traditions and new perspective', *Journal of Heritage Tourism*, 1(1): 1–16.

Tunbridge, J. E. and Ashworth, G. J. (1996). *Dissonant Heritage: The management of the past as a resource in conflict*. Chichester, UK: John Wiley.

UNESCO. (1972). Convention Concerning the Protection of the World Cultural and Natural Heritage. Paris [online]. Available at http://whc.unesco.org/en/conventiontext (accessed 9 January 2012).

UNESCO. (2003). The Convention for the Safeguarding of the Intangible Culture Heritage [online]. Available at www.unesco.org/culture/ich/en/convention (accessed 29 November 2015).

UNESCO. (2008). *Operational Guidelines for the Implementation of the World Heritage Convention*. Paris: United Nations Educational, Scientific and Cultural Organisation.

Wilkinson, D. and Birmingham, P. (2003). *Using Research Instruments*. London: Routledge.

Woon, W. L. and Mui, L. Y. (2010). 'Element cost format for building conservation works in Malaysia', *Structural Survey*, 28(5): 408–19.

Yin, R. K. (2003). *Case Study Research: Design and methods* (3rd edn). London: Sage.

Yin, R. K. (2009). *Case Study Research: Design and methods* (4th edn). London: Sage.

8 Using case-based methods in construction research with complementarities

Fidelis A. Emuze

This chapter promotes an approach for evolving answers to contemporary problems in construction research with the use of an idiographic study as a starting point. The chapter highlights the viewpoint that answers to complex construction research problems often require problem-driven as opposed to method-driven research, because the collaborative use of methods appears to produce answers that can be justified better than those produced by a single method. From a case-based research perspective, the chapter highlights the importance of complementarity with a detailed example. The chapter elaborates on various complementarities between case studies and statistical methods. Through an example, the chapter establishes how a research problem informs the choice of research approach that potentially benefits the need for idiographic analysis, which, in turn, creates a three-dimensional picture of the research questions and answers. The chapter will, therefore, assist readers to position future work in a context where the resolution of a research problem allows common sense and evidences to meet the need for documentation, understanding and statistical validity. The example and discussion in the chapter will encourage readers to employ case-based methods that can best answer a relevant research question.

8.1 Introduction

This chapter discusses case-based methods as approaches that are particularly suitable for evolving evidences that can tackle complexity-related 'wicked problems'. Wicked problems are different from tame problems. Tame problems have clear missions and are aligned with the classical science and engineering model (Hardin, 1968). Tame problems have well-defined, stable limits that can be precisely evaluated (Ritchey, 2013). In contrast, wicked problems are problems that require answers obtained through a multidisciplinary approach (Rittel and Webber, 1973), as such problems tend to be vague and pertain to tough issues that are underpinned by stakeholder dependency (Ritchey, 2013). Wicked problems are said to be in a constant state of flux, as they are sets of complex, interacting issues that are evolving constantly in a social context. In essence, most wicked problems are often without easy technical solutions, because they tend to be malignant and tricky (Ritchey, 2013).

Such problems are exemplified in the range of challenges that must be overcome in construction research. In particular, complexity at the project and industry level has been the subject of much construction research. As an applied discipline, construction research has the dual purpose of solving practical problems and creating theoretical/conceptual knowledge (Azhar *et al.*, 2010). Compared with general management, project management is a practice-oriented field of research with the purpose of improving working 'method/procedure/policy' (Hallgren, 2012). To accomplish this purpose, appropriate research methods are needed. In wicked problems that must be tackled by construction research, there are issues of interdependent causes, wherein the solution to a problem is just one potential cause of an outcome. As mentioned by Blackman *et al.* (2013), as well as Verweij and Gerrits (2013), understanding how interventions interact with the issue and its context requires the use of a method that is able to study how causes combine, rather than independent effects, so as to allow multiple causal pathways, rather than one causal model.

This chapter is aimed at showing that the use of case-based methods and statistical methods in construction research allows complementarity between different methods so as to do justice to the complex nature of construction projects with many actors and stakeholders. The complementarity could bridge the requirements of contextualisation and generalisation, which often contradict each other (Verweij and Gerrits, 2013). To realise this aim, Section 8.2 presents problem-driven research and highlights the areas of complementarity between case studies and statistical methods. Section 8.3 provides an exposition on the rationale for case-based methods. The comparative strengths of the combined methods are also discussed, before qualitative comparative analysis (QCA) as a case-based method is introduced. Section 8.4 provides an example of QCA in construction, and Section 8.5 concludes the chapter.

8.2 Problem-driven research in construction

Ways of doing research often rely on theory, method and phenomenon. Although several texts use either question-driven or phenomenon-driven research, this chapter uses the term problem-driven research (PDR). When articles in leading databases are examined, the dearth of PDR is apparent in many fields (Markusen, 2015), including applied fields such as construction management. This has encouraged a theoretical straightjacket in academic research. The obligation placed on a research in terms of the identification of a gap in extant theory has contributed considerably to the perpetuation of this straightjacket (Ferris *et al.*, 2012). The theoretical straightjacket is a figurative representation of the way that theory and a focus on theory have imposed progress limits on a field of study (Schwarz and Stensaker, 2014). In acknowledging the issue of relying excessively on theory-driven research (TDR), Schwarz and Stensaker (2014) argue that PDR provides a platform for moving from incremental advancements in theory to the generation of new and bold knowledge. PDR starts with a question about practice-oriented phenomena and it strives to answer the question with reasoning, evidence and methods that would yield the best responses to the question (Markusen, 2015). The problems described in a PDR are significant, and the design should explore causality, explanation and description (Markusen, 2015).

To explain the nature of PDR, Table 8.1 summarises its properties and refers to the use of an approach that can lead to the discovery of knowledge that can alter practice and policy. A PDR goes beyond the extension of an extant theory to solving practical problems that may have multiple causations. For example, if a policy change is the goal, a PDR should state the policy concerns at the outset of the project. Table 8.1 alludes to the position that, in a good PDR, the problems should be posed, their importance should be emphasised, and the methods should be chosen for their power to evolve answers. Thus, a research that is done with the situations summarised in Table 8.1 should yield insights that are useful to organisations and policymakers who are keen on the improvement of phenomena plaguing construction practice.

In contrast to PDR, methods-driven research starts by proposing, improving or applying new methods to phenomena to produce extensions and results that are not pre-arranged for a specific practical problem (Markusen, 2015). However, Sternberg (2007) argues that adopting this route has inherent flip sides. The flip sides include the following:

- Answers are often not forthcoming for questions that are not susceptible to the methodological fad of the time.
- Researchers who want to answer such questions find themselves marginalised.
- A field comes to view questions as important to the extent that they conform to particular methods, instead of the other way around.
- Training becomes increasingly narrow to the extent of creating a self-perpetuating monopoly of the preferred method that is hard to break.
- When the fad breaks, researchers find themselves unprepared to switch gears because they have not been trained to think or do research in a broad set of ways that will prepare them for emergent trends.

Table 8.1 Properties of problem-driven research

Aspect	PDR-based descriptor
Aim of research	Contributions to knowledge and the facilitation of conventional understanding (for practice)
Motivation for research	Understand a managerial or organisational phenomenon that has a problem or multiple problems; use the comprehension of the phenomenon for capturing and extending knowledge
How the contribution is made	The contribution is made by mapping existing or new constructs on to a phenomenon
The role of theory	Empirical data are used to position or build theory, and multiple theories are used to describe and explain phenomena
Primary target audience	Academics and practitioners in the industry
Research output	Radical advancement of current knowledge through development of new theories or ideas; allows the extension and/or new combination of existing theories

Source: Adapted from Schwarz and Stensaker, 2014: 486

Therefore, Sternberg (2007) contends that research should be problem-driven as opposed to method-driven (or theory-driven – emphasis on theory from the author of this chapter). The importance of elevating PDR is underscored by the contention of Markusen (2015) that, even in the field of management, there is recognition of the need to move across disciplinary fields from a narrow TDR approach to alternative views on what constitutes a contribution in a research endeavour. However, the view expressed in this chapter is that PDR should enjoy equal status with other approaches to an enquiry. According to Sternberg (2007), the use of PDR can be encouraged in three ways. First, a research must be evaluated on the basis of the importance of the questions it asks, rather than on the fit of the questions to the research paradigms in vogue. Second, the research is subjected to further evaluation on the basis of how well the questions posed are answered. Third, there has to be acknowledgement of the genuine value of research, which is the discovery of new know-how that can assist society to better understand phenomena through the verification or falsification of existing theories. As Wachtel (1980, cited in Sternberg, 2007: 38) rightly opined, 'we will thrive best when we encourage investigators to use the methods that best suit them, as well as the more general approaches to research that best suit them'. This quote emulates the call for methodological pluralism in construction management research by Professor A. R. J. Dainty (Dainty, 2008).

In order to tackle research problems/questions in the construction sector, therefore, a range of methodologies is used. For example, Table 8.2 shows that a broad range of methods is in use within the construction engineering and management (CEM) research community. Taylor and Jaselskis (2010) indicate that within the 1,102 articles that were published in the *Journal of Construction Engineering and Management* (*JCEM*), between 1993 and 2007, experimental, survey, qualitative and quantitative case, non-empirical theory building and simulation can be found. The observation shows the multiplicity of methodologies in CEM.

The observation supports the view that construction research is an applied field that embraces methodological pluralism to solve research questions. At the core of these methodologies is the question of what kind of understanding the researchers should be trying to achieve based on the problem to be solved (Rooke et al., 1997).

Table 8.2 Research methodologies employed in *JCEM* (1993–2007)

Research methodology	Use frequency (%)
Experimental research	27.0
Survey research	22.0
Qualitative and quantitative case research	16.0
Non-empirical theory building research	13.0
Simulation research	12.0
Optimisation research	9.0
Archival data set research	< 1.0

Source: Taylor and Jaselskis (2010: 1)

Case-based methods offer the opportunity to explore diverse kinds of understanding, with the use of the complementarity that exists between case studies and statistical methods (Flyvbjerg, 2011). Mixed methods research is now gaining traction, as scholars are beginning to see the different methodological approaches in a complementary light (Creswell and Plano Clark, 2011). For these scholars, research is problem-driven, and methods are employed to best answer the research question (Flyvbjerg, 2011). By leveraging more often on the strengths and weaknesses of different methods, say case studies and statistical methods, such scholars are able to locate best answers to a problematic and/or complex question: a wicked problem. Complementarities between case-study methods and statistical methods are summarised in Tables 8.3 and 8.4. The main strength of statistical methods is 'breadth', in the form of large-N studies, which allows understanding of the extent of a phenomenon within a given population. Another major strength of statistical methods is precision, which fosters testability and reliability that can be achieved in variables (Meredith, 1998). Another notable strength relates to the knowledge and wide acceptance (and replication) of their standard research procedure. This procedure includes model formulation, variance reduction and sample-size selection (Meredith, 1998).

These strengths of statistical methods are evident in construction management research that is rooted in the positivist paradigm (Dainty, 2008). Incidentally, these strengths of statistical methods can compensate for the weaknesses of case studies, as illustrated in Table 8.3. According to Flyvbjerg (2011), Meredith (1998) and other notable scholars, such as Yin (2014), the major strength of the case study is depth in the form of detail, richness, completeness and within-case variance. Other major strengths of case studies include significant conceptual validity, understanding of context and the associated process and understanding of the root causes of phenomena and their related outcomes. Most importantly, case studies foster new hypotheses and research questions (Table 8.4). Meredith (1998) observes that, in case-study methods:

1   a phenomenon can be studied in its natural setting, so that meaning and relevant theory can be generated from the understanding gained by observing actual practice;
2   the much more meaningful question of why, rather than what and how, can be answered with a full understanding of the nature and complexity of the whole phenomenon;
3   where the variables are still unknown and the phenomenon is not well understood, exploratory research can be done;
4   richness of explanations and its potential for testing hypotheses in well-described, specific situations is possible.

These strengths of case-study methods compensate for the weaknesses that are associated with statistical methods. Notable weaknesses of statistical methods include sampling difficulties, which, in turn, appear to limit empirical generalisations. In particular, the statistical method is often associated with objective measures of variables, which may lead to trivial results that do not go beyond the original model of the study. In effect, when data are collected out of context, they cannot account for crucial variation related to context and they tend to produce reliable, but insignificant, 'so what' results (Meredith, 1998). However, supporting hard empirical data with illustrative case studies is useful, as the studied cases provide solid examples of abstract concepts and processes (Fellows and Liu, 2008).

The strengths and weaknesses shown in Tables 8.3 and 8.4 can be illustrated as follows. For example, if a scholar wants to understand a phenomenon in any level of detail, say rework and cost overrun in construction (please see rework-related articles by Professor P. E. D. Love and associates: Love and Edwards, 2004; Love *et al.*, 2008), in terms of what causes it and how to prevent it, such a scholar needs

Table 8.3 Complementarity: Strengths of statistical methods versus weaknesses of case-study methods

Strengths of statistical methods	Weaknesses of case-based methods
Breadth/Large N sample	Small N sample owing to access or time
Understanding the spread of a phenomenon within a population	Weak understanding of phenomenon occurrence in a population of study
Measures of correlation for population of cases	Selection bias may overstate or understate relations
Establishment of probabilistic levels of confidence/testability	Statistical significance is always either unknown or unclear
Standard procedure	Unfamiliarity of procedures
Precision	Need for multiple methods and tools
Reliability	Lack of controls

Source: Adapted from Flyvbjerg, 2011: 314, and Meredith, 1998: 443

Table 8.4 Complementarity: Strengths of case-study methods versus weaknesses of statistical methods

Strengths of case-based methods	Weaknesses of statistical methods
Depth of information/data	Trivial data and thin results
High conceptual validity	Conceptual stretching, through grouping together dissimilar cases to get larger samples
Understanding of context and process	Weak understanding of context, process and causal mechanisms
Understanding of what causes a phenomenon, and linking of causes to outcomes	Correlation does not imply causation
Fostering of new hypotheses and new research questions	Weak mechanism for fostering new hypotheses
Relevance	Weak relevance may come about owing to sampling difficulties, limited model, low explained variance and variable restrictions

Source: Adapted from Flyvbjerg, 2011: 314, and Meredith, 1998: 443

to do case studies (see Love *et al.*, 2008). If the scholar is interested in understanding how widespread the rework is, how it correlates with other design and construction problems and varies across different projects, subsectors and industries, and at what level of statistical significance, the scholar should conduct a study that employs statistical methods (see Love and Edwards, 2004). However, if the scholar wants to understand both, which is advisable if the scholar would like to speak with relative authority about a problem (such as the drivers of conflict in infrastructure projects sponsored by donor agencies), the scholar must conduct case studies and statistical methods (see Schaffer-Boudet *et al.*, 2011). Thus, the core of the discussion of this chapter applies to scholars who would like to gain breadth and depth of knowledge in relation to phenomena of interest. Both depth and breadth of knowledge are crucial to finding solutions to practical problems, and the acceptance of complexity and its recognition in research through integrated processes and products should engender a rigorous use of methodological pluralism in construction research (Fellows, 2010). If construction management researchers are effectively to solve the problems that the industry faces, they have to adopt methodological approaches that allow them fully to understand phenomena that influence business and project performance in construction (Love *et al.*, 2002).

8.3 The 'why' and 'how' of case-based research

Creswell (2013) explains the qualitative case study (QCS) as the exploration of a 'bounded system' over time through detailed, in-depth data collection involving multiple sources of information, each with its own sampling, data-collection and -analysis strategies. The outcomes of the QCS are the description and explanation of cases, which involve case-based themes. Thus:

> Case studies are analyses of persons, events, decisions, periods, projects, policies, and other institutional systems that are studied holistically with the use of one or more methods. The case that is the subject of the inquiry will be an instance of a class of phenomena that provides an analytical frame – an object – within which the study is conducted and which the case illuminates and explicates.
>
> (Thomas, 2011: 23)

Philosophical thinking associated with case-based methods

Research philosophy is about the development and nature of knowledge (Collins, 2010). It involves ontology, epistemology, axiology and methodology, apart from paradigms that provide the lens through which the world is viewed (Collins, 2010). Thus, a paradigm is a conceptual framework for how reality is seen (for example, positivism), so as to determine what it is like through its basic constituents (ontology), and what the nature and status of knowledge (epistemology) are (Silverman, 2013). As shown in Table 8.5, pragmatism is the interpretive framework associated with case-based methods/case studies. According to Creswell (2013), in practice, the researcher using pragmatism will use multiple methods of data

Table 8.5 Philosophical beliefs for case-based methods

Interpretive framework	Ontology – nature of reality	Epistemology – how reality is known	Axiology – role of value	Methodology – enquiry approach
Pragmatism	Reality is what is useful, what is practical and what works	Reality is known through deduction, induction and abduction; through the use of many research tools, reality is inferred from evidence	Values are derived from the way that knowledge indicates the perceptions of researchers and participants	The collection of data and analysis in the research process includes both qualitative and quantitative approaches

Source: Adapted and modified from Creswell, 2013: 37

collection to best answer the research question, will collect data from multiple sources, will focus on the practical implications of the enquiry and will endeavour to emphasise the significance of conducting the research that best answers the research question. Based on this description of pragmatism, case-study researchers have consistently made use of both quantitative and qualitative data. For instance, *The Case Study Anthology*, edited by Yin (2004), shows many examples of pragmatic considerations in research. In the edited anthology, the editor clearly highlights methodological issues involved in each chapter.

In case-based methods, pragmatism puts aside the quantitative/qualitative divide and rejects the paradigm debate to provide a platform for researchers in terms of locating the answers to research questions (Feilzer, 2010). The aim is to think about cases where a proper and explicit dialectical synthesis can be done between causes and interpretations to engender explanations (Byrne, 2013). Given the following, it is important to understand and discuss the characteristics of case-based methods in the next section:

> Paradigms concern ontology and epistemology and so, relate directly to methodology and, thence, proceed to data collection and analyses. Here, it is essential to understand the terms clearly and to express the position adopted. All should be justified from theoretical and (or) pragmatic considerations – self-awareness to inform the research.
>
> (Fellows, 2010: 11)

Characteristics aligned with case-based methods

The threefold logic of case studies comprises outcome/process selection, contextualisation and process tracing (Mjøset, 2013). To realise any of these logics in construction, a case study is a detailed study of actors, organisations, teams, events and projects. As far back as 2001, Flyvbjerg (2001) argued for the explanatory power of case-based studies, especially in relation to the formulation of policy interventions. In fact, Byrne (2013) contends that systematic comparison case studies are the foundations of useful theoretical descriptions of the social world. The data for case studies are thus derived from observations, interviews, questionnaires and archival notes, as case studies are used as sources of insights and ideas to describe phenomena, to apply concepts and theories in an effort to understand the management of projects, and to illustrate principles in construction research (Fellows and Liu, 2008).

These definitions and descriptions of case studies are reflected in how researchers have employed case-based methods. The study of individual cases and the methods for understanding them are clearly case-based (Kent, 2013). As an illustration, Table 8.6 shows that detail is an essential part of case studies, and the unit of analysis in the construction research context may vary from an individual to project teams. Although not shown in the table, it is important to note that the discipline background of case studies draws from many related disciplines, and the structure for reporting case studies depends on the convention adopted by the relevant organisation. Regardless of the style adopted by a reporting organisation, it is important for the nature of reporting to indicate the purpose of the study. Such a purpose could be explanatory, descriptive or exploratory. For example, the type of compositional structure for descriptive studies could be any of linear-analytic, comparative or chronological (Yin, 2014).

The nature of problems suitable for case-based designs should necessitate in-depth description and analysis of phenomena, so that proper causality and meanings can emerge from the study. This is in tandem with Ragin (1997), who contends that case-oriented researchers see cases as meaningful, but complex, configurations of events and structures. Such researchers treat cases as singular, and entities are

Table 8.6 Characteristics of case-study approaches

Characteristic	Descriptor
Focus/aim/target	Developing an in-depth description and analysis of a case or multiple cases
Nature of problem	Required in-depth understanding of phenomena through cases
Unit of analysis	Detailed study of an individual, a team, a task, series of tasks, a work method, a project, a programme, an organisation, an event, series of events, etc.
Data collection	Multiple sources, through interviews, surveys, documents/archival records/databases, various forms of observation, various forms of artefact, etc.
Data analysis	Multiple strategies that include pattern matching, explanation building, time-series analysis, logic models and cross-case synthesis
Nature of report	Detailed analysis of one or more cases with linear-analytic, comparative, chronology, theory-building, suspense or non-sequenced structure used for composition

Sources: Creswell, 2013; Yin, 2014

purposively selected. The researchers refrain from the use of identical observations drawn at random from a pool of equally likely selections. Table 8.7 shows the criteria that may be used in the selection of cases (Bennett and Elman, 2007; Flyvbjerg, 2011). The descriptors in the table show the perceived attributes of cases, and these attributes influence the decision of the researcher in terms of selection.

Authors such as Seawright and Gerring (2008) have also produced a detailed exposition of the techniques required for the selection of cases in QCS. Seawright and Gerring (2008) describe the case-selection procedures for typical, diverse, extreme, deviant, influential, most-similar and most-dissimilar cases. The description addressed large-N techniques, their uses and their representativeness.

According to Ragin (1997), purposive (information-oriented) selection of cases is aligned with case-based research. Information-oriented selection is always done to maximise the utility of information from small samples and single cases. Cases are selected on the basis of expectations about their information content (Flyvbjerg, 2011). Bennett and Elman (2007) also mention that it is crucial that cases be selected purposively so as to maximise inferential leverage. Citing several authors, Bennett and Elman (2007) went further, to say that the selection criteria that are applicable

Table 8.7 Strategies for information-oriented selection of cases

Criterion	Descriptor
Deviant cases	To obtain information on unusual cases, which can be especially problematic or good in a more closely defined sense; to understand the limits of existing theories and to develop new concepts, variables and theories that are able to account for extreme cases; deviant cases do not conform to the predictions made by a theory or theories under investigation
Maximum-variation cases/most-similar and least-similar cases	To obtain information about the significance of various circumstances for case process and outcome; for example, between three and five cases that are very different on one dimension: size, location, budget, etc.
Critical cases/ least-likely	To obtain information that permits logical deductions of the type, 'if this is (not) valid for this case, then it applies to all (no) cases'; the least-likely case study relies on an inference that follows a Bayesian logic: the more surprising an outcome is relative to extant theories, the more confidence is increased in the theory or theories that are consistent with such an outcome; in other words, if the theory can make it here, it can make it anywhere
Paradigmatic cases	To develop a metaphor or establish a school of thought (intellectual tradition) for the domain that the case concerns
Combining cross-case and over-time comparisons	This design can generate considerable inferential leverage from the study of a few cases that combine cross-case and over-time (or before–after) case comparisons; this allows each case to be potentially compared in two different ways: the before of Case A at time T^0 can be compared to the after of Case A at time T^1; also, Case A in one or both periods can be compared with another Case B, which may be divided into two periods

Sources: Adapted from Flyvbjerg, 2011, and Bennett and Elman, 2007

to large-N studies are not always useful for qualitative purposes. There are other rationales for studying particular kinds of case, as shown in Table 8.7. Ragin (1997) further suggests that case-based studies start with the idea that social phenomena in like settings (such as organisations, cities, countries, regions, cultures and so on) may empirically parallel each other sufficiently to permit their comparison and contrast. Thus, case-based researchers often must strive to justify their choice of cases and the scope of their arguments. The scope often sets empirical and theoretical limits on the extent of the generalisation that can be made from the results of a study. Both case-study methods and statistical methods require a statement of scope at the start of the study.

Scope is closely related to generalisation. Scope also influences the understanding of causation in a case-based study, either in simple causal explanations or separate causal paths to the same outcome (Goertz and Mahoney, 2013). Selection of cases and the scope of a study are items that must be addressed when a study is designed. The extant literature, in particular social science, includes a plethora of research designs that combine case-study methods with statistical methods. Combining methods is said to be a powerful approach, as each method can, to a large extent, offset the limitations of other methods. As a case in point, researchers have used case studies either to test or to demonstrate the operation of hypothesised mechanisms (Bennett and Elman, 2007). Indeed, the narrative in Bennett and Elman (2007) shows that combining statistical methods and case studies gives a more comprehensive insight into a phenomenon. The use of case studies side by side with statistical methods can help to identify and clarify causal mechanisms, point out measurement problems in statistical methods, identify omitted variables, highlight interactions effects, indicate when unit heterogeneity exists and recognise different systems and types of a major phenomenon, such as a study on site fights in pipeline projects located in different countries (McAdam *et al.*, 2010). Byrne (2013) went further, to note that case-based methods are needed to:

- clarify causation and to specify the range of applicability of a stipulated account of a causal mechanism;
- clarify causes beyond the unique specific instance (the object of idiographic enquiry);
- specify the limits of generalisation;
- engender proper social scientific understanding that is associated with complexity theory and critical realism – in other words, it addresses complexity;
- promote explicit dialectical synthesis between cause and meaning/interpretation in order to achieve explanation.

These arguments by Byrne (2013) are corroborated in the infrastructure research of Verweij and Gerrits (2013). Case studies in construction research may view the case as an analysis of interactions among project actors in the procurement of a particular project. Interviews, hard documentary evidence and questionnaire surveys may be used to obtain information relevant to a case so as to produce deep insights, which may be narrow in breadth (Fellows and Liu, 2008). In brief, case studies endeavour to answer the how and why questions by focusing on contemporary events (Yin, 2014).

Comparative strengths of combined case-study–statistical work

The work of Bennett (2002) is apt in illustrating the advantages of combined case-study and statistical methods. Bennett (2002) notes that an interesting use of case-study–statistical work is the investigative development and testing of theories. Within the realm of operations research, Meredith (1998) also mentions that this method is preferred for building new operational management theories. Bennett (2002) also contends that the combination of methods allows the use of statistical analysis to identify deviant cases, and also the use of case-study analysis to develop new or left-out variables that can then be added back into the statistical analysis.

With regard to the testing of theory, there is a strong consensus concerning the ways in which each method can help offset the limitations of the other. Referring to several authors, Bennett (2002) affirms that statistical analysis can identify generalisations across populations, and case studies can guard against the potential inaccuracy of statistical correlations by using *process tracing* to examine in detail the causal mechanism hypothesised to lie behind statistical correlations. Flyvbjerg (2011) provides an example of process tracing in action when the falsification of optimism as a global explanation of executive failure was addressed. In addition, case studies can more readily address problems related to interaction, certain selection effects and complex variable over large-N cases. In contrast, criticism of the combined case-study–statistical method is that the combination is not consistently precise on the matter of selecting cases from a population. Rather, different case-selection criteria are necessary for different theory-building purposes (see Table 8.7). Nevertheless, the authors of *The SAGE Handbook of Case-Based Methods* (Byrne and Ragin, 2013) allude to the perspective that:

- case-based methods are useful and represent a way of moving beyond a narrow tradition in research (for example, construction research) that has set quantitative and qualitative modes of exploration, interpretation and explanation against each other;
- generalisation is best understood as involving careful attention to the scope of a study;
- given the uniqueness of construction outputs, either projects or products, and the context-specific nature of location, universal laws are impossible; however, comparative case-based methods would allow scholars to give accounts beyond a unique idiographic range, and individual case studies can contribute to this; the scoping requirement is to pay attention to the limitations of researchers' knowledge claims in time and space.

The aforesaid perspectives suggest that case-based methods offer the opportunity to see complementarity in action through the use of comparative methods. Comparative methods, in their different forms, are central to case-based understanding and are always seen in terms of classification on the basis of comparison and the exploration of complex causalities (Byrne, 2013). In this framework, classification compares the separation of like from not like. Systematic comparison based on interpretation of a range of cases seeks to establish typical features of particular cases or sets of cases and to explore how those features taken together are causal to the prevailing state of cases.

Comparative case-based methods in a nutshell

A case-based method could adopt either quantitative *and* qualitative strategies or either one of the two. The conceptualisation of case-based methods explicitly rejects a fundamental distinction between the quantitative and qualitative (Byrne, 2013). The approach to case studies in construction research has the scope to be either positivist or intrepretivist (Fellows and Liu, 2008). Case-based methods are, however, not limited to small-*N* research alone, but could be large-*N* studies, where the limitations and assumptions of mainstream quantitative methods limit the conclusions from statistical analysis. It is notable that evidence from leading journals in the social sciences and the management disciplines suggests that the use of process tracing (Hall, 2013; Rohlfing, 2013), nested analysis (Lieberman, 2005) and QCA (Ragin *et al.*, 2003) is prevalent in terms of causation analysis within comparative research. A similar search in leading built environment journals, however, shows that only QCA is making a limited inroad in construction research. As a result, a short overview of QCA is provided in the next section, before an example of QCA in construction research is presented. The example illustrates complementarity and not the actual QCA steps within the studies.

Qualitative comparative analysis

Kent (2013) mentions that the classification and comparison of several cases to determine similarities and differences and to explore complex causality are case-based. Case-based methods include configurational analysis, correspondence analysis and cluster analysis. Configurational analysis with QCA, which was developed by Ragin (1987), is used in sociology, political science, management and economics (Jordan *et al.*, 2011; Rihoux *et al.*, 2013). Ragin (1987) proposes that researchers adapt the techniques of Boolean algebra to inventory permutations of many dichotomous variables and to compare more cases than qualitative researchers normally do. Ragin (2000, 2008) developed QCA for the exploration of causality among cases. The method enables researchers to recap outcomes in many cases, while considering the way each independent variable depends on context. The case-based method (rather than variable-based approach) is informed by ideas of QCS, which incorporate systematic cross-case comparison.

QCA uses a reasoning based on the membership of sets and logical reduction of conditions until only certain conditions that clearly differentiate between outcomes are included in explanations (Blackman *et al.*, 2013). The results that are derived from this procedure represent causal pathways to outcomes. Blackman *et al.*

(2013) show that QCA works with a particular conception of causality based on multiple conjunctural causations. This means that outcomes may emerge from distinctive combinations of causes that are indicated in types of case, instead of independent variable effects. Whereas statistical methods assume that a certain condition always takes effect in the same direction, QCA, in contrast, presumes that the effect of a condition depends on context (Sager and Andereggen, 2012). Methodologically, QCA provides a bridge between qualitative and statistical methods, as it integrates the features of case-based and variable-based research (Ragin, 1987). QCA builds on Boolean algebra principles, which allow researchers to tackle 'complex and seemingly contradictory patterns of causations', while also removing irrelevant causes (Ragin, 1987: 118). As a result, Blackman *et al.* (2013) propose that QCA is a non-linear, non-additive and non-probabilistic conception that stresses the diversion of complex combinations of conditions and different paths that lead to the same outcome. In other words, QCA, as a technique for exploring the complementarity between case studies and statistical methods, provides a basis for the qualitative consideration of complex policy problems based on '*like-with-like*' comparisons between cases. Although very limited QCA work can be found in the construction research literature, Jordan *et al.* (2011) provide guidance on its use in construction, write about QCA applications in construction and also make suggestions to scholars employing it in construction.

8.4 QCA in construction research: DBFM project

To highlight the application and use of QCA in a practical context, an empirical example is drawn from an article published in the *International Journal of Project Management* by Verweij (2015). The article presents how managers in a public–private partnership (PPP) project responded to social and physical events, which shaped project outcomes. The study examines the events of the implementation of a Dutch design, build, finance and maintain (DBFM) highway infrastructure project. The research question for the study asked: 'How do managers in the implementation phase respond to events occurring in the context of PPP transportation infrastructure projects, and which management responses produce satisfactory outcomes?'

Rationale for the research approach

The reported properties of the problem to be solved influenced the chosen approach of the empirical study. In particular, context and complexity in PPP infrastructure project management show that a bridge between qualitative and quantitative methods of research is needed. The article demonstrates the fact that infrastructure projects are implemented in a complex sociophysical (sociotechnical) context. It recognises that projects are open systems that are contextually embedded in open systems. The author also premises the reason for the study on issues that include:

- failure of variable-oriented studies to explain how good outcomes are actually produced when compared with case-based methods in relation to the style that is most beneficial for realising outcome of complex decision-making processes;
- the realisation that a transportation infrastructure project can be understood to consist of strings of cases, which entail different combination of similar elements.

This aforesaid realisation led to the conclusion that cases within a project exhibit similarities and differences. The research is therefore a '*problem-driven*' research.

Research approach

Based on the exposition of the nature of the research problem, Verweij (2015) used multi-value QCA (mvQCA), a variant of QCA, for the study. The research approach consists of four steps. Step 1 employed grounded collection of data with the use of open qualitative interviews. Twenty interviews were conducted, and the interview data were supported by the information obtained through site visits, document reviews and website inspections. Step 2 textually coded the transcribed interview data for events and employed the use of Atlas.ti for the process. The application of the mvQCA method was Step 3. The mvQCA method that was used followed the usual four routines of QCA analysis, and Tosmana software was used in this instance. The fourth step is the interpretation of the analysed results. Twenty cases were studied in the research.

Salient result and implications

The findings show that, in most cases, for the highway DBFM project, social events that are responded to by managers with externally oriented actions produce satisfactory outcomes, whereas internally oriented management responses are related to unsatisfactory outcomes. Verweij (2015) contends that an implication of the study, in the Netherlands, is that construction firms are yet to fully master the ability to manage relationships with local stakeholders that comes with the DBFM type of PPP. The author, therefore, suggests that, in procuring future DBFM projects, public managers should recognise the stakeholder capabilities of all project parties. This suggestion is implementable by relevant parties to such projects, and, as such, it has the attribute of a practical wisdom.

8.5 Reflection on the method in relation to complementarity

The narrative in the article clearly shows areas in which trade-offs were made between the advantages and disadvantages of both case-study methods and statistical methods. In specific terms, complementarity that may have been derived by the use of the mvQCA method includes:

- breadth and depth of insights that arose through detailed, rich and complete study of the project in the form of multiple cases as configurations of events, conditions and outcomes;
- high conceptual validity, as the findings resonate with previous work, especially in terms of project management – importance of the principal's involvement in the implementation phase of a DBFM project;
- effective understanding of the context and process, which informs the design of the empirical work, among other decisions;
- understanding of complex causation and linking causes to outcomes, as evident in all the tables in the article: data matrix, truth table, etc.

8.6 Summary

This chapter discusses case-based research in terms of its suitability for construction research. As a problem that has multiple pathways, complexity should be addressed with methods that can best answer a construction research question. The chapter, therefore, highlights the relevance of problem-driven research to outcomes that seek to solve practical quagmires in the industry. Although theory-driven and method-driven studies flourish in the literature, based on their level of usage by researchers, the properties of problem-driven research align it to an approach that would be appropriate to tackle complexity and context issues in construction. These properties of problem-driven research imply that a researcher should properly deliberate methods chosen for a study that is practice-focused. The chapter thus argues for pragmatism as a research philosophy, and the use of case-based methods, which include statistical methods so that valuable insights are related to phenomena, emerges through complementarity to strengthen the robustness of an enquiry.

The chapter also shows the influence that case selection and scope have on the extent of generalisation that would come out of a study. Although limited use of case-based methods with statistical methods appears to exist in construction, extant literature is inundated with their theory and practice. Therefore, it can be argued that extant methodological texts have helped to clarify best practices pertaining to diverse QCS approaches so much that exact criteria are beginning to emerge. The continuing interest in mixed methods has resulted in a clearer understanding of the comparative strengths and weaknesses of different methods. The offsetting of weaknesses with corresponding strengths offered by another method within a combination of methods promises a rigorous employment of case-based research in construction. Within the construction management field, which is renowned for positivism, and research that sought answers to complicated issues, QCS could serve a useful purpose.

As illustrated by the example in this chapter, construction research can replicate the widespread benefits of case-based methods in the discipline. The limited use of combined case-study methods and statistical methods in construction research is notable. The use of the two approaches could provide a platform for emergent insights in construction research. The combination of these methods seems to be useful for studies that explicitly address multiple causal pathways, context and generalisation. The bridge between the two methods is complementarity, which enables the leveraging of the strengths of both methods. A study that seeks depth and breadth of know-how about a phenomenon could use case-based methods, after due consideration and thought have been expended on the contextualisation of the research question. A study that would be relevant to both academia and industry would benefit from a methodological standpoint that allows a problem-driven enquiry.

Two areas constitute challenges that should be overcome. First, although the larger social science and management literature has good examples of case-based methods research, very little has been written in the construction context. This invariably leads to limited general techniques and approaches for combining case-study methods and statistical methods that are available to case-based researchers in the applied discipline. Although there are many possible ways to combine case studies and statistical methods, the complementarity and challenges of the sequence of work deserve further examination. Second, the use of case-based methods with statistical methods cannot proceed without new related skills and knowledge that are specific to construction research. This would come about through the evolution of a critical mass of case-based research. QCA, which is gradually gaining traction in the field, could serve as a starting point in this regard. In creating awareness and encouraging the use of case-based methods, researchers should endeavour to open their minds to existing work in social and management sciences, where steps and examples of case-based methods abound. It is, however, important to note that the use of case-based methods with statistical methods requires a multi-year research project, so as to gain meaningful problem and methodological insights from the exercise.

Notes

1   The author wishes to thank Stefan Verweij and colleagues, who prefer to be anonymous, for their help in improving an earlier version of this chapter.
2   Interested construction researchers, especially case-based researchers who are keen on exploring new ways of tackling project complexities, should make an effort to read the entire articles from Verweij (2015) and Schaffer-Boudet *et al.* (2011).
3   For an informative approach to the use of QCA in construction-related studies, please see Jordan *et al.* (2011), which comprehensively addresses it.
4   For detailed insights about the techniques for case selection within a case-based research approach, please see the work of Seawright and Gerring (2008).

References

Azhar, S., Ahmad, I. and Sein, M. K. (2010). 'Action research as a proactive research method for construction engineering and management', *Journal of Construction Engineering & Management*, 136(1): 87–98.

Bennett, A. (2002). 'Where the model frequently meets the road: Combining statistical, formal, and case study methods', *Proceedings of the American Political Science Association Conference* (pp. 1–27). Boston, MA, 28–31 August 2002.

Bennett, A. and Elman, C. (2007). 'Case study methods in the international relations subfield', *Comparative Political Studies*, 40(2): 170–95.

Blackman, T., Wistow, J. and Byrne, D. (2013). 'Using qualitative comparative analysis to understand complex policy problems', *Evaluation*, 19(2): 126–40.

Byrne, D. (2013). Case-based methods: Why we need them; what they are; how to do them. In D. Byrne and C. C. Ragin (eds), *The SAGE Handbook of Case-Based Methods* (pp. 15–38). London: Sage.

Byrne, D. and Ragin, C. C. (eds) (2013). *The SAGE Handbook of Case-Based Methods*. London: Sage.

Collins, H. (2010). *Creative Research: The theory and practice of research for the creative industries*. Lausanne, Switzerland: AVA.

Creswell, J. W. (2013). *Qualitative Inquiry and Research Design: Choosing among five approaches* (3rd edn). Los Angeles, CA: Sage.

Creswell, J. W. and Plano Clark, V. L. (2011). *Designing and Conducting Mixed Methods Research* (2nd edn). Los Angeles, CA: Sage.

Dainty, A. R. J. (2008). Methodological pluralism in construction management research. In A. Knight and L. Ruddock (eds), *Advanced Research Methods in the Built Environment* (pp. 1–11). Oxford, UK: Wiley-Blackwell.

Feilzer, M. V. (2010). 'Doing mixed methods research pragmatically: Implications for the rediscovery of pragmatism as a research paradigm', *Journal of Mixed Methods Research*, 4(1): 6–16.

Fellows, R. (2010). 'New research paradigms in the built environment', *Construction Innovation*, 10(1): 5–13.

Fellows, R. and Liu, A. (2008). *Research Methods for Construction* (3rd edn). Chichester, UK: Wiley-Blackwell.

Ferris, G. R., Hochwarter, W. A. and Buckley, M. R. (2012). 'Theory in the organisational sciences', *Organisational Psychology Review*, 2(1): 94–106.

Flyvbjerg, B. (2001). *Making Social Science Matter*. Cambridge, UK: Cambridge University Press.

Flyvbjerg, B. (2011). Case study. In N. K. Denzin and Y. S. Lincoln (eds), *The SAGE Handbook of Qualitative Research* (pp. 301–16). Thousand Oaks, CA: Sage.

Goertz, G. and Mahoney, J. (2013). Scope in case study research. In D. Byrne and C. C. Ragin (eds), *The SAGE Handbook of Case-Based Methods* (pp. 307–17). London: Sage.

Hall, P. A. (2013). 'Tracing the progress of process tracing', *European Political Science*, 12(1): 20–30.

Hallgren, M. (2012). 'The construction of research questions in project management', *International Journal of Project Management*, 30(7): 804–16.

Hardin, G. (1968). 'The tragedy of the commons', *Science*, 162(3859): 1243–8.

Jordan, E., Gross, M. E., Javernick-Will, A. N. and Garvin, M. J. (2011). 'Use and misuse of qualitative comparative analysis', *Construction Management & Economics*, 29(11): 1159–73.

Kent, R. (2013). Case-centred methods and quantitative analysis. In D. Byrne and C. C. Ragin (eds), *The SAGE Handbook of Case-Based Methods* (pp. 184–207). London: Sage.

Lieberman, E. S. (2005). 'Nested analysis as a mixed-method strategy for comparative research', *The American Political Science Review*, 99(3): 435–52.

Love, P. E. D. and Edwards, D. J. (2004). 'Determinants of rework in building construction projects', *Engineering, Construction & Architectural Management*, 11(4): 259–74.

Love, P. E. D., Edwards, D. J. and Irani, Z. (2008). 'Forensic project management: An exploratory examination of the causal behavior of design-induced rework', *IEEE Transactions on Engineering Management*, 55(2): 234–47.

Love, P. E. D., Holt, G. D. and Heng, L. (2002). 'Triangulation in construction management research', *Engineering, Construction, & Architectural Management*, 9(4): 294–303.

McAdam, D., Schaffer-Boudet, H., Davis, J., Orr, R. J., Scott, W. R. and Levitt, R. E. (2010). 'Site fights: Explaining opposition to pipeline projects in the developing world', *Sociological Forum*, 25(3): 401–27.

Markusen, A. (2015). 'Problem-driven research in regional science', *International Regional Science Review*, 38(1): 3–29.

Meredith, J. (1998). 'Building operations management theory through case and field research', *Journal of Operations Management*, 16(6): 441–54.

Mjøset, L. (2013). The contextualist approach to social science methodology. In D. Byrne and C. C. Ragin (eds), *The SAGE Handbook of Case-Based Methods* (pp. 15–38). London: Sage.

Ragin, C. C. (1987). *The Comparative Method: Moving beyond qualitative and quantitative strategies*. Los Angeles, CA: University of California Press.

Ragin, C. C. (1997). 'Turning the tables: How case-oriented research challenges variable-oriented research', *Comparative Social Research*, 16(1): 27–42.

Ragin, C. C. (2000). *Fuzzy-Set Social Science*. London: University of Chicago Press.

Ragin, C. C. (2008). *Redesigning Social Inquiry: Fuzzy sets and beyond*. London: University of Chicago Press.

Ragin, C. C., Shiman, D., Weinberg, A. and Gran, B. (2003). 'Complexity, generality, and qualitative comparative analysis', *Field Methods*, 15(4): 323–40.

Rihoux, B., Álamos-Concha, P., Bol, D., Marx, A. and Rezsöhazy, I. (2013). 'From niche to mainstream method? A comprehensive mapping of QCA applications in journal articles from 1984 to 2011', *Political Research Quarterly*, 66(1): 175–84.

Ritchey, T. (2013). 'Wicked problems: Modelling social messes with morphological analysis', *Acta Morphologica Generalis*, 2(1): 1–8.

Rittel, H. W. J. and Webber, M. M. (1973). 'Dilemmas in a general theory of planning', *Policy Sciences*, 4(2): 155–69.

Rohlfing, I. (2013). 'Varieties of process tracing and ways to answer why-questions', *European Political Science*, 12(1): 31–9.

Rooke, J., Seymour, D. and Crook, D. (1997). 'Preserving methodological consistency: A reply to Raftery, McGeorge and Walters', *Construction Management & Economics*, 15(5): 491–4.

Sager, F. and Andereggen, C. (2012). 'Dealing with complex causality in realist synthesis: The promise of qualitative comparative analysis', *American Journal of Evaluation*, 33(1): 60–78.

Schaffer-Boudet, H., Jayasundera, D. C. and Davis, J. (2011). 'Drivers of conflict in developing country infrastructure projects: Experience from the water and pipeline sectors', *Journal of Construction Engineering & Management*, 137(7): 498–511.

Schwarz, G. and Stensaker, I. (2014). 'Time to take off the theoretical straightjacket and (re)introduce phenomenon-driven research', *The Journal of Applied Behavioral Science*, 50(4): 478–501.

Seawright, J. and Gerring, J. (2008). 'Case selection techniques in case study research', *Political Research Quarterly*, 61(2): 294–308.

Silverman, D. (2013). *Doing Qualitative Research* (4th edn). London: Sage.

Sternberg, R. J. (2007). 'The importance of problem-driven research: Bringing Wachtel's argument into the present', *Applied & Preventive Psychology*, 12(1): 37–8.

Taylor, J. E. and Jaselskis, E. J. (2010). 'Introduction to the special issue on research methodologies in construction engineering and management', *ASCE Journal of Construction Engineering & Management*, 136(1): 1–2.

Thomas, G. (2011). *How to Do Your Case Study: A guide for students and researchers*. London: Sage.

Verweij, S. (2015). 'Achieving satisfaction when implementing PPP transportation infrastructure projects: A qualitative comparative analysis of the A15 highway DBFM project', *International Journal of Project Management*, 33(1): 189–200.

Verweij, S. and Gerrits, L. M. (2013). 'Understanding and researching complexity with qualitative comparative analysis: Evaluating transportation infrastructure projects', *Evaluation*, 19(1): 40–55.

Yin, R. (ed.) (2004). *The Case Study Anthology*. Thousand Oaks, CA: Sage.

Yin, R. (2014). *Case-Study Research: Design and methods* (5th edn). Thousand Oaks, CA: Sage.

Part IV
Mixed methods research

9 Studentification and the housing needs of international students in Johannesburg

An embedded mixed methods approach

Kola Ijasan and Vian Ahmed

This chapter presents a research that takes a pragmatist philosophy while adopting an embedded mixed method design to determine the existence (or otherwise) of a divergence in the housing needs of international students from those of local students in Johannesburg. The study explored whether international and local students have different accommodation preferences and needs. This was necessitated by the recent increase in the tertiary student population in South Africa and the resultant shortage of adequate residential accommodation for students. This problem has many elements. According to the South African Government Parliamentary Monitoring Group for students' accommodation, one of the main problems is that, in order to provide greater access, accommodation charges are kept low in government-provided residences, but this often leads to poor standards, as revenue often falls below the cost of maintenance. From the private providers' perspective, the main challenge is some students' attitude to public goods with some vandalism and unruly behaviour within the residences: it is claimed that some particular sets of students are notorious for their antisocial behaviour, hence the need for them to control access in some cases. However, from the residents' (i.e. students') point of view, the main problem is the lack of, or unsuitability of, facilities within the residences. Hence, this research focuses on the housing needs of students in student residential accommodation. The findings of the research reveal that there are statistically significant differences between the accommodation needs and preferences of international and local students. There is also a seeming lack of willingness or awareness on the part of developers in this regard.

9.1 Introduction

The population of university students in South Africa increased from about 473,000 in 1993 to approximately 799,700 in 2008 (IEASA, 2011). According to Bara and Celliers (2013), this shows that, within a period of 16 years, the student population almost doubled – the increase equates to 4.3 per cent growth per annum. The number of international students rose to 893,024 by 2010, thereby accounting for almost 8 per cent of the total number of university students. With this rapid growth, the

Ministerial Committee for the Review of Student Housing at South African Universities (Rensburg, 2012) looked into where these students reside and reported that there is a serious shortage of student accommodation at South African universities and in their host communities. It was recorded that the total number of beds at all universities stood at 107,598 in 2010, equating to 12 per cent of the total enrolment. According to the Parliamentary Monitoring Group (PMG) (2013), in South Africa's twenty-three universities, only 5.3 per cent of first-year students – those arguably in the greatest need of accommodation – were in university residences. Out of the total number of contact students across all the universities, only 20 per cent were accommodated at university residences. This research, therefore, looks into the current state of the art regarding students' accommodation in general and international students in particular. It explores their housing needs using a survey in order to see if these needs are being met by housing providers or not.

Many studies have been conducted of the factors that contribute to the experiences of international students in foreign countries (Lacina, 2002; Constantine *et al.*, 2005; Sawir *et al.*, 2008; Appiah, 2011; Ota, 2013). In as much as there are prospects of improvement, international students still face a lot of challenges, and Poyrazli and Grahame (2007) summarised these challenges as follows:

- finding living accommodation, means of obtaining food and other essential items for daily life, and getting appropriate documentation for their stay;
- learning the academic culture, including how to interact with faculty and other students, and different styles of teaching; and
- making new friends and developing a new social support system.

If finding accommodation is one of the key challenges faced by students, it follows by deduction that this could even be more challenging for international students. It is in this light that this work aims to explore the housing needs and preferences of international students in the Braamfontein area of Johannesburg, in order to produce a list of key facilities and attributes needed in residential accommodation that plans to attract international students. This chapter explores the existence of a divergence between the housing needs of international students and those of their local counterparts. The next section reviews key literature on student housing. It introduces the concept of 'studentification', explores students' housing environment and highlights major needs of students in their accommodation. The rest of this chapter will further elaborate on the emerging themes from literature by using multiple sources of data to both identify these needs and also validate the findings.

9.2 Literature review

This section critically evaluates past research into the needs of students in their accommodation and it aims to see if there are any differences in the needs of students with respect to whether they live locally or are from international destinations.

The purpose of this is to explore the theoretical underpinnings of studentification and housing needs and, consequently, put these properly within the context of the South African environment.

Studentification and the student housing market

Student housing is defined as any housing that is solely intended to provide accommodation for students attending college or university, and it creates an environment where social connections, independency and learning to live with people take place (Fogg, 2008). Macintyre (2003) stated that the growing number of students is adversely creating a growing market for students' accommodation, and this trend has continued globally. Universities can either directly, or indirectly through private providers, take all steps to ensure that future student housing projects are conceived as an integral part of the local community. Baron (2011) states that evidence from several countries indicates that, in the last two decades, higher-education students have gradually become a highly influential player in the private rental sector (PRS). Macintyre (2003) indicated that the high demand for student accommodation in some areas has had the effect of drastically driving up the value of residential property in the area. Within the UK context, Rugg *et al.* (2000) highlighted how entrepreneurial landlords and property investors accelerated the supply of private rented student housing to exploit the accommodation needs of students in specific locales. This expanding accommodation need may lead to the concept of '*studentification*'.

Studentification as a term was coined to describe the impact of relatively high numbers of higher-education students moving into established residential neighbourhoods. According to Smith (2005), this process triggers a range of distinct social, economic, cultural and physical transformations within the studentified communities/neighbourhoods. Depending on the current state of the PRS of the community, studentification can have either a positive or negative effect on the host community. Economically, Baron (2011) asserts that:

> The impact of 'studentification' on rent prices can only be evaluated by means of a hedonic price analysis of rental apartment prices. In particular, the impact of 'studentification' can be evaluated by regressing apartment prices on apartment attributes, location amenities and an indicator function identifying apartments that are located in 'studentified' neighbourhoods.

Studying the niche market in the UK in 2002, Rugg *et al.* indicated that tertiary-education students have increasingly become a significant factor in the rental market. The reason why student housing has become increasingly privatised was because of the constantly changing needs of students and their 'expectations with respect to housing and the related amenities' (Niles, 2004). Increasingly, student preference for new and exciting facilities was also seen as a driver for enrolment choices; hence, for colleges eager to attract the best students, 'the quality of student housing and their living environment is a key selling point' (Samuels, 2008).

Students' living environment

According to Crimmin (2008), student housing presents a unique opportunity for student affairs administrators, as they contribute to and also support the educational experience of university students, and also because student housing plays a role in studentification and the students' environment as well. Research by Ware and Miller (1995) concluded that although the lifestyle trends of students differ, student accommodation plays a major role in the accomplishment of students' studies. This is because the facilities that are offered to students can influence them in applying to the university student housing and ensure they keep on living on the student housing campus (Frazier, 2009). A study carried out in Australian universities revealed that students who have safe, steady accommodation often have higher chances of staying in university education than those with less safe, less steady accommodation (University of Adelaide, 2013). Li *et al.* (2005) noted that a major benefit of living on campus is 'higher levels of student satisfaction'. In corroboration, Thomsen and Eikemo (2010) showed that students' odds of reaching graduation are increased by a better living environment, because this leads to a more effective learning environment. Nevertheless, Balogh *et al.* (2005) noted that the majority of the traditional dormitories do not have the highly desired amenities and building infrastructure that students and their parents now see as essential to the university experience in the modern era. Examples of such features include fitness and recreation centres, wireless networking capabilities, kitchens, and bedrooms with private bathrooms. Universities constantly have to compete with privately owned, off-campus accommodation that caters more for the modern-day student. This accommodation offers modern-day amenities and is also close enough to the universities for the convenience factor (Balogh *et al.*, 2005; Muslim *et al.*, 2012).

Borrowing from Crimmin (2008) and Adewunmi *et al.* (2011), key factors that are pertinent in student housing/preference are:

- ability to cook meals;
- flexible lease agreement;
- proximity to university;
- conveniences;
- parking facilities;
- social facilities;
- ability to study within the accommodation.

Studentification is a process evolving from an over-concentration of students in a particular area (usually near university campuses). Over-concentration of students in such neighbourhoods can lead to economic, physical, social and cultural changes based on the students' activities in the neighbourhood (Smith, 2005; Rugg *et al.*, 2010). In areas undergoing studentification, there is a need for accommodation providers, as well as student affairs administrators, to investigate the needs of their tenants and align such needs to the services provided (Samuels, 2008). There are key factors that are pertinent in student housing (Crimmin, 2008; Adewunmi *et al.* 2011).

From the literature reviewed, there seems to be a dearth of research on the needs of international students within the scope of studentification, and that is what this chapter focuses on. This chapter further explores the existence of a divergence in the amenities needed by international students and those needed by local students, and, if such a divergence exists, how developers and student housing providers are reacting to it. The research protocol and underpinning methodological steps adopted are discussed in the following section on research methodology.

9.3 Research methodology

This refers to the combination of different methods used by a researcher to resolve a research problem (see Creswell, 2003; Easterby-Smith *et al.*, 2004; Kumar, 2005; Trochim, 2006). Earlier, Guba and Lincoln (1994) proposed that questions of research methodology are of secondary importance to questions of which paradigm is applicable to the research at hand. Travers (2002), however, mentioned that the methodology of a research should include the researcher's theoretical position and how 'employed methods' have resolved the research question. In a more explicit depiction, Saunders *et al.* (2009) produced the 'research onion', which aimed at summarily explaining all the aspects of a research. Taking its cue from the research onion, this research explains some of the key considerations of the framework and how it informed the methodological choice of this chapter.

Research philosophy

This is the belief about the way in which data about a phenomenon should be gathered, analysed and used. According to Saunders *et al.* (2009), the research philosophy could be positivism, realism, interpretivism, subjectivism, pragmatism, etc. The '*pragmatist*' research philosophy was chosen for this research. This philosophy can be loosely described as being midway between the positivist and interpretivist philosophies (Denscombe, 2008). According to Creswell (2003), 'pragmatist researchers focus on the "what" and "how" of the research problem and they place the research problem as central and apply all approaches to understanding the problem'. Pragmatism is seen as the paradigm that provides the underlying philosophical framework for mixed methods research (Mackenzie and Knipe, 2006). The pragmatist philosophy is particularly suited for this work because it investigates the current state of the housing needs of international students in South Africa and also how the property managers of student accommodation perceive these needs.

Research approach

A researcher has the option of either being inductive in their research – that is, building a theory from specific observation – or being deductive – that is, testing a theory. According to Babbie (2010), whereas deduction begins with an expected pattern that is tested against observations, induction, on the other hand, begins with observations and seeks to find a pattern within them. However, there is another approach to research that deviates from these two. The approach is called *abduction*. Abduction is a research approach that goes from observations to a theory. It aims to find the most likely explanation of the phenomenon being studied. The abductive approach is used in this chapter by, first, presentation of the generic concept of studentification from the literature (with some evidence), then a statistical difference between the needs of local and international students is explored, and, from the available evidence, the conclusions are 'abduced' (Fischer, 2001; Morgan, 2007).

9.4 Research strategy and methodological choices

Research strategy is how a researcher plans to go about answering or addressing a research question. Methodological choice, on the other hand, is about how the researcher hopes to generalise data, that is, qualitatively or quantitatively. According to Saunders *et al.* (2009), the strategies are by experiments, surveys, case studies, etc. This chapter uses surveys as one of the strategies to answer the questions concerning the needs of international students within their residences. When a researcher chooses to view methodology, it should be a holistic combination of research philosophy, choices and method(s) of data collection. The research strategy adopted informs the methodological choice(s) of data collection. The choices available are mono method, mixed method and multi-method. For this chapter, mixed methods were used: initial semi-structured interviews were applied to investigate the nature of the research questions, surveys were later used to explore the needs of the students, and, finally, interviews were used to validate the findings.

As this research is about the exploration of the needs of international students in South Africa and the differences, if any, between these needs and those of local students, the following questions were asked:

- *What data are required?* The data concerned the facilities and conditions needed by international students in their residential accommodation during their study, and how important these facilities were.
- *What methods are going to be used to collect and analyse these data?* Mixed method data-collection techniques were used – that is, literature review, interviews and surveys.
- *How is all of this going to answer your research question?* Two groups of students were surveyed, that is, international and local students. The local students served as a control group. The needs of the international students were analysed quantitatively and compared with those of the local students to see if there was any divergence.

The three-stage data-collection protocol

Stage 1

Five semi-structured interviews were held, with two property managers and three international postgraduate students living within the study area. The purpose of these interviews was to see whether there were additional factors affecting international students' housing needs beyond the scope of what was discovered in the literature. The factors generated from Crimmin (2008) and Adewunmi *et al.* (2011) were further populated within the South African context. The existence of studentification was confirmed, and a potential divergence of housing needs between local and international students was suggested, but, owing to the relatively small number of interviewees, these needs were not yet affirmed.

Stage 2

A questionnaire was developed after the initial interviews and administered to 250 students from the two universities in the study area. Of these questionnaires, 127 were correctly filled in and returned. This made for a response rate of 50.8 per cent. Statistical difference between the needs of local and international students was confirmed at this stage; upon statistical analysis, the research assumptions were either accepted or rejected.

Stage 3

Upon the quantitative data-collection process, three additional interviews were conducted with the property managers of the hostels used. This process helped in cleaning up the data and in contextualising some of the responses. The main purpose of this set of interviews was to validate the findings from the previous two stages. This is further discussed under the analysis section.

9.5 Population, data sampling and unit of analysis

Population in statistics has a specific meaning. It refers to 'all members of a defined group that is being studied'. For this research, the population comprised all international higher-education students in Johannesburg. As we would imagine, it would be burdensome to accurately estimate this group of people; hence the need to get a 'sample' that can represent the population. The method of selecting this sample is called '*sampling technique*'. There are many sampling techniques available to a researcher – for example, random, stratified, non-representative, purposive, etc. As stated earlier, this research aims to investigate the accommodation needs of international students in SA. To achieve this, and to be sure it has been done as accurately as possible, these 'needs' must be compared with those of local students; the local students in this instance serve as a control group. A cross-sectional, non-representative subset of the population was targeted by way of door-to-door data-collection methods in eleven residential hostels within the Braam area. Three research students helped distribute the same online questionnaire to pre-identified student hostels within the study area. This method is a form of *purposive sampling*. For the quantitative data, the unit of analysis – that is, the entity studied – comprised the individual students surveyed, whereas the unit for the main interviews comprised the hostels whose managers were interviewed.

9.6 Data analysis and interpretation

Stage 1: Semi-structured interviews

After research ethics approval had been obtained, five people were involved in the interview exercise. Two property managers were interviewed in depth, with three international PhD students. All interviews were conducted between 6 May and 19 May 2014. Initial interviews were helpful in populating the factors pertinent in student accommodation facilities and also in contextualising some of these factors. The nature of the seven factors from the literature review was operationalised, and localised meanings were given to these factors. Besides helping to collect data, the interviews assisted in developing and testing the adequacy of other proposed research instruments and assessing the effectiveness of the research methods proposed. All interviews were held at the participants' choice of location and time. Interviews took 45–60 minutes; they were recorded, and member-checking protocol was followed to ensure reliability (see Lincoln and Guba, 1985). The following section articulates some of the key lessons learned from the interview exercise. A summary of key foci is presented below.

Focus 1: Evidence of studentification

It was accepted that the Braamfontein area of Johannesburg is being studentified. It was noted that even students whose family reside in other parts of Johannesburg prefer to move to Braam for many reasons, one of which is the proximity to both the universities: Witwatersrand and Johannesburg. It was seen that, as suggested by Smith (2005), these increases in students' activities have also made rental prices increase, and this has subsequently led to gentrification within the area. There is also an increasing cost of student accommodation in the area and the virtual shutdown of business activities during school holidays. This research considered a list of facilities developed by Adewunmi *et al.* (2011) within a Nigerian university context and by Crimmin (2008) within a US environment; it was realised that these factors might have different connotations for students in South Africa. For example, 'proximity to university' is not a function of distance from the university but rather a matter of whether the accommodation is close to the routes of bus services commonly provided by the universities. Also, flexibility of lease arrangements is more concerned with the possibility for the students to pay rent for 11 months and not for the January holiday month. In all, it can be seen that the initial interviews helped significantly in providing a better understanding of the situation, reduced the ambiguity of the questionnaire and provided a smooth transition to the main survey. The participants helped, not just in aligning the process in preparation for the main survey, but also in providing some respondents. Upon completion of the initial semi-structured interview exercise, the researchers distributed 250 questionnaires and received 127 back. The responses are presented in the next section.

Focus 2: How to ensure participation of international students in this study

As observed from the literature and personal experiences, most international students in South Africa are postgraduates (mostly PhD candidates), and this set of students does not usually have classes or regular meeting venues. This begged the question of how to locate them for participation in this survey. Suggestions from the PhD students were that research seminar venues could be potential recruitment grounds, as most attendees of these events are postgraduate students. Also, the two managers of the residential hostels kindly agreed to distribute and retrieve twenty copies of the questionnaire for the research.

Focus 3: Appropriateness of the research questions and questionnaire structure

In the assessment of the survey tool, three key lessons were learned. Another purpose of this process was to test the appropriateness of the questionnaire being developed. The interviewees were shown draft copies of the questionnaires for comments and inputs. Going into the exercise, two sets of similar questionnaires were designed for local and international students separately. First, the PhD students suggested that a better structure would be to consolidate the two sets of questionnaires into one and make a distinction (i.e. local or international student) in the profile section of the survey; second, the proposed supposition regarding the gender of the respondents was excluded from the final draft. Unlike Teshome (2010), it was argued that there wouldn't be any need to explore the needs discrepancy based on gender, as many of the respondents might be living with their partners, and their needs would be more joint needs rather than gender-dependent needs. Finally, the number of units within their hostel was seen to be a potentially difficult question to answer, because most of the halls are designed in such a way that they all have private entrances. This means that there is an element of privacy built into most of the accommodation.

Stage 2: Survey and results (statistical analysis of findings)

This section contains the analysis and findings of the survey research. The questionnaire survey was developed from scratch and it incorporated the key housing needs already identified by Crimmin (2008) and Adewunmi *et al.* (2011). There were a number of areas that were evaluated within the questionnaire; however, owing to the scope of the research, this section focuses on selected areas of the research problem. The study carried out shows the differences in the accommodation needs and preferences of international and local students. Through the data collection and statistical analysis, the accommodation needs and preferences of international and local students were compared. From our findings, we gauged the importance of certain factors to students when they selected their accommodation. Furthermore, we evaluated the difference in this importance between local and international students.

Profile of students and accommodation features

There were fifty-eight respondents aged between 25 and 35, and sixty-nine respondents aged between 35 and 45. These age groups made up the majority of the respondents within our study. There were eighty-nine female and twenty-eight male respondents. The majority of the respondents were studying at postgraduate level, with 76 per cent of the total respondents coming from this group. Seventy-five (59.2 per cent) students lived in Braamfontein, twenty-eight (22.1 per cent) in Parktown, four (3.1 per cent) in Hillbrow, four (3.1 per cent) in Melville and sixteen (12.4 per cent) in other parts of the city. As the survey was conducted around July, it was seen that 89 per cent of the respondents had lived in their current accommodation for 6 months or more. Table 9.1 shows some other features of the accommodation types and facilities. More international students than local students lived in private accommodation, no international student lived in a dormitory, and the two respondents that chose 'other' specified that they lived with family. The section of the questionnaire on the number of people sharing accommodation was designed to explore the issue of privacy within the students' residential spaces. Table 9.1 presents the average responses from all the students (i.e. local and international). As shown, based on forty-one responses for local students, on average, six students shared a kitchen/kitchenette, whereas the figure was 2.2 for international students, based on eighty-six responses. The facility least shared by international students was the bedroom, at 1.3 on average. This difference could be attributed to the fact that local students might have known each other for some time or come from the same neighbourhoods, and, hence, it would be easier for them to share spaces and save costs, whereas international students often come to the country alone, or with their spouse, and so the chances of meeting perceived strangers and sharing intimate spaces with them are avoided as much as possible.

Table 9.1 Students' profiles and accommodation features

Accommodation provider	Local students	International students
State	0	0
Private (off campus)	12	58
University (on campus)	29	28
Accommodation type		
Apartment/flats	20	67
Dormitory	3	
Communal living	18	17
Others	0	2
No. of people sharing accommodation facilities		
Bedroom	1.9	1.3
Kitchen(ette)	6	2.2
Bathroom	4.3	2.6
Living area	6	2.2
Garden/balcony area	8.5	3.2

The needs of the students were subdivided into two – that is, needs within the students' rooms and needs around the accommodation. Table 9.2 shows the needs of the surveyed students within their room. It is seen that the local students have fewer (pending) needs, as only room heaters and microwave ovens were flagged as preferred needs that hadn't been met. On the part of the international students, many more of their needs were yet to be met by their accommodation providers. Noise insulation ranked highest, with fifty-four respondents, and adequate storage followed closely at fifty-one.

Table 9.3 shows the second class of students' needs – that is, the needs within the compound. The needs of these two sets of students again appear to be divergent, as shown in Table 9.2. However, the need for the provision of sports/gym facilities was common to both sets of students. The facility least requested by international students was a 'car wash'. This can be explained by the fact that international students don't readily buy cars as soon as they arrive in the country.

Table 9.2 Housing needs within the respondents' rooms

Facility	Local students (41)		International students (86)	
	Provided	*Preferred*	*Provided*	*Preferred*
Study desk/table	32	6	39	43
Bed	38	3	54	24
Bedding	31	9	41	32
TV/radio	19	12	32	43
WiFi ADSL	34	4	37	47
Bookshelves	30	8	39	43
Reading lamp	32	4	51	27
Bedroom cupboard/storage	34	2	32	51
Curtains/blind	32	4	34	21
Refrigerator	37	3	41	43
Room cooler, e.g. fan, AC etc.	29	11	30	17
Room heater	17	21	43	35
Smoke extractor	29	11	23	28
Fire alarm	32	4	34	11
Cooking stove	35	5	45	27
Iron/ironing board	26	12	43	21
Kettle	23	11	38	44
Microwave oven	23	15	41	37
Crockery, e.g. spoons, forks etc.	22	5	43	16
Cookware, e.g. pots and pans	22	5	29	43
Noise insulation	28	11	29	54
Telephone and extension services	34	2	34	28
En-suite bath/shower, toilet	34	7	34	50
Wash basin	31	9	23	49
Wall clock	4	2	7	5

Table 9.3 Housing needs within the respondents' compound/immediate surroundings

Facility	Within the compound			
	Local students		International students	
	Provided	Preferred	Provided	Preferred
Free transport to place of study	36	5	37	39
Computer/PC suite	34	2	29	14
Laundry facility	31	5	29	38
Mail facilities	30	4	36	12
Study room	27	11	32	21
Integrated dining plan	6	5	7	18
Gym/sports facility	5	34	37	34
Secure parking	4	37	39	11
Prayer room	4	11	2	17
Games/TV room	6	35	11	17
Incorporated utilities bill	4	31	39	23
Access control	38	2	41	28
Car-wash area	16	21	23	9

Assumption testing

Assumption 1 of this research was that there is no statistically significant difference in the accommodation needs, choice based on features, usage of accommodation areas by student, and security and safety of accommodation of local and international students. As there are just two groups of students – 'international and local' students – a *t*-test analysis was conducted. The purpose of this *t*-test was to inferentially add a higher level of significance to the earlier descriptive analysis of Tables 9.2 and 9.3. The summary of the output is presented in Table 9.4.

Table 9.4 shows there is a significant difference in the accommodation needs ($t(125) = -22.06$; $\rho = 0.00$), choice based on features ($t(125) = -31.77$; $\rho = 0.00$), usage of accommodation area by students ($t(125) = 21.11$; $\rho = 0.00$), and security and safety of accommodation ($t(125) = 35.96$; $\rho = 0.00$) of local and international students. This implies that the assumption was wrong.

Assumption 2 was that there was no correlation between accommodation provider and available facilities. To analyse this, a Pearson correlation was conducted. Pearson's *r* correlation coefficient is a statistical measure of the strength of a linear relationship between paired data. It is denoted by *r* and depicted as $-1 \geq r \leq 1$. A positive *r* value means a positive linear correlation between the tested variables and vice versa. The correlation table is presented in Table 9.5.

Table 9.5 shows that there was a significant negative correlation ($r = -0.26$, $\rho = 0.00$) between the accommodation providers and the facilities available in their accommodation. This assumption is, therefore, wrong.

In summary, the survey revealed that local students shared more facilities than international students; international students seemed to want spaces to themselves

Table 9.4 Independent sample *t*-test of accommodation

	Student type	Mean	SD	T	df	p	Remark
Needs	Local	61.24	9.88	-22.059	125	0.00	Significant
	International	102.16	9.72				
Features	Local	33.78	6.94	-31.768	125	0.00	Significant
	International	75.98	7.03				
Usage area	Local	10.71	4.87	-21.109	125	0.00	Significant
	International	30.74	5.06				
Safety and security	Local	17.98	3.22	-35.959	125	0.00	Significant
	International	37.84	2.75				

Note: The p-value measures how significant an independent variable is to the response variable; when $p \leq 0.05$, it denotes a statistical significance between the variables measured

and have facilities for their sole use – communal living was more prevalent with local students. This is further buttressed by the inferential analysis of 'needed' facilities versus 'provided' ones.

Stage 3: Response contextualisation interviews

The concluding process of the research was to validate the results of the survey by conducting follow-up interviews. Three interviews were conducted, in September 2014, with property managers of three of the residences where questionnaires were distributed. All interviewees were university graduates with a minimum of 5 years' residential property management. The combined portfolio of the three companies was estimated at 5 billion rand. For the purpose of confidentiality in this analysis, they are referred to as R 1, R 2 and R 3.

The aim of this round of the exercise was to get the property/facilities manager's perspective of the situation and to offer contextual interpretations of some of the statistical results. The interviews took an average of 76 minutes to complete and were conducted at the hostel premises. Owing to the minimal number of interviewees, the authors used a thematic method of content analysis. Thematic analysis,

Table 9.5 Correlation of accommodation provider and available facilities

	N	r	p	Remark
Accommodation provider* available facilities	127	-0.26**	0.00	Significant

Note: The p-value measures how significant an independent variable is to the response variable; when $p \leq 0.05$, it denotes a statistical significance between the variables measured

according to Braun and Clarke (2006), is a qualitative analytic method for 'identifying, analysing and reporting patterns (themes) within data. It minimally organises and describes your data set in (rich) detail'.

Having identified that there is a shortage of residential spaces in the country and a large concentration of university students in Johannesburg, this research surveyed 250 students and received 127 valid responses. It is understood that this number does not represent all students in the municipality; however, the sampled students serve as a window into the current situation and, hence, provide a glimpse of the bigger picture. This section gives a brief breakdown of some of the views of the managers in relation to the survey findings. These are presented in two themes, as shown in the next section.

Accommodation preferences

It wasn't a surprise that most of the respondents lived off campus: the bias was from the areas on which the survey concentrated. Also, all three interviewees weren't surprised that most of the international students lived in apartments and flats. R 2 and R 3 mentioned that they were sure that, for the local students, the availability and quality of alternative residences is one of the factors considered when choosing universities. This corroborates Samuels (2008). Previous stages of data collection did not focus on reasons for the students' preferences; however, this round shed more light on why international students would prefer to live in flats and apartments rather than town houses. It was proposed that a possible reason for this would be for security purposes.

Most international students are mature students and they loathe the idea of communal living or shared facilities. R 1 noted that the types of social interaction of mature international students are different from those of local students. This is in line with the findings of Lacina (2002), who noted that, often, international students want to see more of the country rather than spend time with fellow house mates. This, he said, is evident, given the fact that the house managers are mostly called for simple tasks and questions that could have been resolved by the person next door. R 2 said that they give a questionnaire to potential students asking them about their income and their housing budget. This questionnaire showed that foreign students and, indeed, workers (young professionals) always allocate higher percentages of their income to housing than local residents. According to him, 'if or not they are been truthful with such percentages is another question'.

It was seen that, even when international students have to move to small(er)/shared rooms owing to lack of choice at the time of arrival in the country, they always demand bigger/private rooms as soon as such spaces become available. According to R 3:

> The locals usually have family and friends over every other weekend or so, one would have expected them to be the ones with preferences for private spaces, this is however not the case, I guess it is the African sharing spirit.

To summarise this theme, it can be said that the quantitative findings are not out of place, as most of the results are close to what the interviewees expected.

Housing needs

The housing needs results were tested, and the general consensus from all the respondents was that they take feedback and suggestions from their tenants seriously and they try their best to meet their demands. However, owing to financial costs and the emergence of slum landlords in the inner cities, property managers are not able to be selective in the services they offer. R 2 noted that, in one of the flats in Braamfontein, they offer full furnishing as an option: 'the problem comes when residents start settling and start mixing and matching. They start saying what furniture they want and not, where then do we store the excess?'. For this reason, they offer basic furnishing at a standard price for the entire duration of the lease, without option for any removal. On the point of TV/radio and ADSL being a common need, R 1 pointed out that the reason for this could be the fact that South Africa has eleven national languages, and most non-South Africans would find the local TV unsuitable for them; hence, they need services such as DSTV (which is a premium cable service).

Bookshelves were thought by all respondents to be standard fittings within their flats; there was a mixed feeling as to why international students pointed these out as one of the facilities they were missing. Noise insulation and en-suite baths and toilets are closely tied to the issue of privacy and the seeming length of time it takes for the international students to fully settle into their new environment. This need was also linked to the local students wanting entertainment spaces such as TV rooms more than international students. According to R 3, this has little to do with being studious: it is owing to the fact that local students are more likely to have made friends and, hence, they want to watch their favourite shows with them. R 3 further added that they tried putting 27-inch TVs into their rooms a few years ago, but the insurance company would not allow that, as it was deemed to be too tempting (theft-wise) and a potential source of disruption.

Within the housing needs theme, there was a divergence of opinions on the issue of integrated-dining facilities for residents. According to R 1, the problem is not the lack of capacity to make meals for the students: they know that most students are primarily concerned with studying, and any assistance, especially with food, is a big boost. However, the issues here, according to them, are optimisation and timing. For optimisation, R 1 explained that there is a refectory in the hostel, but, with more than 300 residents, it is difficult to cook without waste. That was what the management of the building decided whenever the issue was brought up. However, they didn't want to convert the refectory into a public restaurant. The second issue is similar to the first and has to do with timing. He said that international students might want the food service for the first couple of months after their arrival in the country, but, as soon as they are settled, they would want to stop; this results in a form of loss for them. With R 2, the main issue is the diverse food needs of international students, and he didn't think it was a good idea for them, as 37 per cent of all their residents are more than 30 years of age. He thought the integrated-food option is best for first-year local students who are leaving home for the first time.

The third respondent, for her part, reckoned that property developers and managers would see it as a good idea. She suggested further research into it and to see how that potential could be maximised. Going further, she surmised that there is a future for this in their residences and that outsourcing it to the right partner would be key in achieving success in that venture.

9.7 Conclusion

In summary, the research has highlighted the key housing needs and preferences of international students in Johannesburg. It shows that there is 'studentification' of the Braam area of Johannesburg with the continued increase in the number of students in this area. It has shown statistically that the needs of local and international students are divergent. These findings have been further corroborated by qualitative findings. Although the results of the findings are not generalisable, given that the conclusions are drawn from a limited set of observations, they are, however, significant indicators of the factors affecting the housing needs of international students in Johannesburg. It was observed that the questionnaire used had a limitation, which was that it should have explored the issue of ease of communication with the property managers. Two of the interviewees noted that it would be interesting to see how the residents view communication ease with their property managers, as they opined that some of the things ticked as 'preferred needs' could be items that the property manager(s) could have provided with discretion.

References

Adewunmi, Y., Omirin, M., Famuyiwa, F. and Farinloye, O. (2011). 'Post-occupancy evaluation of postgraduate hostel facilities', *Facilities*, 29(3/4): 149–68.

Appiah, E. K. (2011). 'Factors that impact international students' learning of introductory physics at Georgia State University'. Unpublished thesis, Georgia State University [online]. Available at http://scholarworks.gsu.edu/phy_astr_theses/13 (accessed 1 December 2015).

Babbie, E. (2010). *The Practice of Social Research* (12th edn). Belmont, CA: Cengage Learning Wadsworth.

Balogh, C., Grimm, J. and Hardy, K. (2005). 'ACUHO-I construction and renovation data: The latest trends in housing construction and renovation', *Journal of College & University Student Housing*, 33(2): 51–6.

Bara, M. and Celliers, C. (2013). 'Travel patterns and challenges experienced by University of Johannesburg off-campus students', *Journal of Transport & Supply Chain Management*, (1): 7.

Baron, M. G. (2011). The impact of studentification on the rental housing market. Haifa: s.n. Available at www-sre.wu.ac.at/ersa/ersaconfs/ersa10/ERSA2010finalpaper204.pdf (accessed 14 January 2015).

Braun, V. and Clarke, V. (2006). 'Using thematic analysis in psychology', *Qualitative Research in Psychology*, 3(2): 77–101.

Constantine, M. G., Anderson, G. M., Berkel, L. A., Caldwell, L. D. and Utsey, S. O. (2005). 'Examining the cultural adjustment experiences of African international college students: A qualitative analysis', *Journal of Counseling Psychology*, 52(1): 3–13.

Creswell, J. W. (2003). *Research Design: Qualitative, quantitative, and mixed methods approaches* (2nd edn). Thousand Oaks, CA: Sage.

Crimmin, N. P. (2008). *An Evaluation of College Sophomore Living Environments: Traditional residence compared to a living learning community with respect to interaction with faculty, peers, and satisfaction with living area*. Dissertation, Johnson & Wales University, Rhode Island.

Denscombe, M. (2008). 'Communities of practice', *Journal of Mixed Methods Research*, 2: 270–83.

Easterby-Smith, M., Thorpe, R. and Lowe, A. (2004). *Management Research: An introduction*. London, Sage.

Fischer, H. R. (2001). 'Abductive reasoning as a way of world making', *Foundations of Science*, 6(4): 361–83.

Fogg, P. (2008). 'Dorm therapy: A design psychologist helps colleges create spaces where students live and learn' [online], *The Chronicle of Higher Education*, 54(26): B24. Available at http://chronicle.com.turing.library.northwestern.edu/weekly/v54/i26/26b02401.htm (accessed 4 June 2014).

Frazier, W. R. (2009). *A Study of Themed Residential Learning Communities at a Midwest Four-Year University: North Dakota State University*. PhD dissertation 3362697, North Dakota State University. ProQuest Dissertations & Theses (PQDT) database. Available at ttp://search.proquest.com/docview/304961600?accountid=42518.

Guba, E. G. and Lincoln, Y. S. (1994). Competing paradigms in qualitative research. In N. K. Denzin and Y. S. Lincoln (eds), *Handbook of Qualitative Research* (pp. 105–17). Thousand Oaks, CA: Sage.

International Education Association of South Africa (IEASA). (2011). International Students Trends in South Africa [Online]. Available at www.ieasa.studysa.org/resources/study_sa_11/international-students-trends-in-south-africa.pdf (accessed 21 April 2014).

Kumar, R. (2005). *Research Methodology: A step-by-step guide for beginners* (2nd edn). London: Sage.

Lacina, J. G. (2002). 'Preparing international students for a successful social experience in higher education', *New Directions for Higher Education*, 117: 21–7.

Li, Y., Mack, C., Shelley, I. and Whalen, D. (2005). 'Contributors to residence student retention: Why do students choose to leave or stay?', *Journal of College & University Students Housing*, 32(2): 28–36.

Lincoln, Y. and Guba, E. (1985). *Naturalistic Inquiry*. Beverly Hills, CA: Sage.

Macintyre, C. (2003). 'New models of student housing and their impact on local communities', *Journal of Higher Education Policy & Management*, 25(2): 109–18.

Mackenzie, N. and Knipe, S. (2006). 'Research dilemmas: Paradigms, methods, and methodology', *Issues in Educational Research*, 16(2): 193–205.

Morgan, D. L. (2007). 'Paradigms lost and pragmatism regained', *Journal of Mixed Methods Research*, 1: 48–76.

Muslim, H. M., Karim, A. H. and Abdullah, I. C. (2012). 'Satisfaction of students' living environment between on-campus and off-campus settings: A conceptual overview', *Social & Behavioral Sciences*, 68: 601–14.

Niles, S. (2004). 'Student housing privatization: A valuable option for addressing student housing shortages, increased student enrollments, reduced school budgets and changing student needs', *Property Writes*, 8(1).

Ota, A. (2013). *Factors Influencing Social, Cultural, and Academic Transitions of Chinese International ESL Students in U.S. Higher Education*. EdD thesis, Portland State University. Available at http://pdxscholar.library.pdx.edu/cgi/viewcontent.cgi?article=2050&context=open_access_etds (accessed 6 December 2015).

Parliamentary Monitoring Group (PMG) (2013). Meeting of the Higher Education and Training, Equity Index in South African Universities: Briefing by Deputy Minister of Higher Education and Training and the Transformation Oversight Committee [online]. Available at https://pmg.org.za/committee-meeting/16621/ (accessed 3 November 2014).

Poyrazli, S. and Grahame, K. M. (2007). 'Barriers to adjustment: Needs of international students within a semi-urban campus community', *Journal of Instructional Psychology*, 34: 28–45.

Rensburg, I. (2012). Report on the Ministerial Committee for the Review of the Provision of Student Housing at South African Universities. South Africa: Department of Higher Education and Training.

Rugg, J., Rhodes, D. and Jones, A. (2000). *The Nature and Impact of Student Demand on Housing Markets*. York, UK: York Publishing Services.

Rugg, J., Rhodes, D. and Jones, A. (2002). 'Studying a niche market: UK students and private-rented sector', *Housing Studies*, 17(2): 289–303.

Samuels, S. (2008). Leveraging Core Competencies: Partnerships in Student Housing. Master of Science in Education Program Northwestern University [online]. Available at www.sesp.northwestern.edu/docs/masters/975226526494ad3687f283.pdf (Accessed 12 October 2014).

Saunders, M., Lewis, P. and Thornhill, A. (2009). *Research Methods for Business Students* (5th edn). Harlow, UK: Pearson Education.

Sawir, E., Marginson, S., Deumert, A., Nyland, C. and Ramia, G. (2008). 'Loneliness and international students: An Australian study', *Journal of Studies in International Education*, 12(2): 148–80.

Smith, D. P. (2005). 'Studentification': the gentrification factory? In R. Atkinson and B. Bridge (eds), *Gentrification in a Global Context: A new urban colonialism?* London: Routledge.

Teshome, Y. (2010). 'Social and institutional factors affecting the daily experiences of the spouses of international students: Voices from the Midwest and implications to academic institutions'. Unpublished doctoral dissertation, University of Iowa.

Thomsen, J. and Eikemo, T. A. (2010). 'Aspects of student housing satisfaction: A quantitative study', *Journal of Housing & the Built Environment*, 25(3): 273–93.

Travers, M. (2002). *Qualitative Research Through Case Studies*. London: Sage.

Trochim, W. M. (2006). Research methods knowledge base, Drake University [online]. Available at www.socialresearchmethods.net/kb/intreval.htm (accessed 11 April 2014).

University of Adelaide. (2013). *2012 Annual Report* [online]. Adelaide, SA: The University of Adelaide. Available at www.adelaide.edu.au/publications/pdfs/a-report-12.pdf (accessed 24 August 2015).

Ware, T. E. and Miller, M. T. (1995). *Current Research Trends in Residential Life*. Tucaloosa, AL: University of Alabama.

10 Enabling building-information-modelling-capable small and medium enterprises

Aisha Abuelmaatti and Vian Ahmed

It is widely recognised that collaboration in the architecture, engineering and construction (AEC) sector increases productivity and improves quality, competitiveness and profitability. This poses the problem of what are the most appropriate collaboration methods available to accommodate unique work settings and virtual-organisation-like modus operandi. With building information modelling (BIM) being imposed by the government on AEC projects by 2016, the development of new technology for collaboration purposes is one solution to this problem. Unfortunately, information technologies have failed in productivity growth, despite previous governments' attempts – most famously, with the 2004 Latham and 2008 Egan reports. BIM is not about a specific technology, but about ensuring the whole-sector adoption is fundamental. In fact, many SMEs within the sector would like to be implementing BIM, not only for the business productivity, but also so as not to get left behind. It is, however, commonly observed that SMEs have fewer chances to get it right, and so they need guidance. Debate over whether or not to use BIM has become almost irrelevant. Instead, the question raised is how collaborative technologies promote collaborative working between large companies and SMEs in the AEC sector to increase productivity and improve quality to meet the high collaboration demands? Against this background, several approaches from the literature were reviewed. This led to identification of the key elements on which to focus during the collaborative-technologies implementation to enhance successful collaboration. A qualitative study of fourteen cases was developed in order to map the current practice of the collaborative-technologies implementations and their success level in the AEC sector. The case studies revealed a number of factors affecting the success of collaborative-technologies implementations. These factors were affecting the collaboration at the organisation level. This also led to further investigation of the factors affecting the collaboration at the end-user level. The quantitative study of a survey returned by sixty-four respondents was guided by the list of elements proposed by the review developed from the literature. The results suggested that the success of collaborative technologies depends on twenty-three factors. This research concluded that collaborative technologies are effective methods to support collaboration within the AEC and meet the demands of the sector, when strategically implemented, underlined by a set of factors. Recommendations that emerged from these findings are used to suggest a path for SMEs to implement BIM into AEC, so that they are better prepared for the future ways of working.

10.1 Introduction

The architecture engineering and construction (AEC) sector has a continuous demand for collaboration owing to the multidisciplinary nature of this sector. In fact, for the highly information-dependent and cost-conscious AEC sector, which was and still is synonymous with delay, waste and inefficiency, the opportunity to use IT is too good to miss. This is encouraged by the 2011 Government Construction Strategy, stating that the AEC sector 'has not fully taken advantage of the full potential offered by digital technology' (Cabinet Office, 2011). This detailed programme aims to reduce the cost in the sector's activities by up to 20 per cent. One of the notable objectives within this document is the requirement of using full collaborative BIM as a minimum for projects by 2016 (Whyte *et al.*, 2011).

Previous research shows that large AEC companies adopting emerging collaborative technologies frequently fail to achieve the full benefits from their implementations (Gladwell, 2001; Baldwin, 2004; Wilkinson, 2005; e-Business W@tch, 2007; Rezgui, 2011). Unlike other technologies, collaborative technologies are very much concerned with collaboration across the project life cycle, and their successful implementation, therefore, will not only require a state of readiness within one company but also within all companies involved in the project life cycle. The general picture of AEC is of a sector that is a pyramid, with control being in the hands of large players and a large base of SMEs. It ensues, naturally, that SMEs are key players in supporting the large companies. This suggests that only through developing a clear understanding of SMEs' needs can effective collaboration through the use of technologies take place. This highlights the importance of recognising how end-users collaborate.

It is commonly observed that SMEs are likely to magnify the sector trend and to be less technically forward-thinking than large companies. Although an extensive body of literature exists on the topic of technological delivery by AEC companies, it tends to concentrate on big businesses, and the experience of SMEs in this area has gone largely unreported. Although there are some notable exceptions (Acar *et al.*, 2005; Abbott *et al.*, 2006; Manley, 2006, 2008), Barker (2012) quoted David Saffin, senior partner at Consulting Engineer SME, which uses BIM: 'smaller firms have fewer chances to get it right and so need guidance to chart a clear path to implementation'. As such, SMEs need to undergo dramatic changes in order to keep up with a changing sector. According to Philip (2012), 'ensuring SME whole-sector adoption of BIM is fundamental to the success of the Government Construction Strategy's 2016 BIM objective'.

Literature reviewed suggests that the organisational and end-user dimensions are very important for the success of collaborative technologies, and that technical characteristics are rarely the reason for the failure of collaborative-technologies implementation. Below is a summary of the key points gained from the literature reviewed that are relevant to how collaborative technologies promote collaborative working in the AEC sector to increase productivity and improve quality to meet the high collaboration demands:

- *Decision-making and perception in relation to change*: Unlike other technologies, collaborative technologies are very much concerned with collaboration across the project life cycle, and their successful implementation, therefore, will require a state of readiness, not only within one company, but also within all companies involved in the project life cycle. Neither small nor medium-sized AEC companies could be considered close to e-ready. A large gap exists between large business ideas and concepts on one hand, and practical implementation in SMEs, on the other.

- *Process vision development*: A classification of business processes can be divided into four groups: core, support, management and business network. Technology push alone, even though to some extent still dominant in AEC, will not lead to competitive advantage. The extent to which organisations can successfully use IT to improve their business or achieve competitive advantage depends on the level of their maturity in managing business process improvement and in managing and utilising their IT infrastructure. Process management describes the future state of process and, therefore, links business strategies with procedures and actions.

- *Strategic implementation adoption*: IT can only be used effectively for a business if it is utilised as part of the information systems (IS) strategy that reflects work practices, people and information. Understanding IT is not equivalent to understanding IS. Technical staff usually focus on the technology, leaving the identification of business needs (process of information) to end-users. This hampers communication between IT professionals and end-users and is a major factor in the failure of many systems.

- *Amending structure*: Collaborative technologies can potentially bring about improvements and offer strategic advantage to organisations if IS/IT strategies are aligned with business strategies. The focus of the implementation shifts to ensure that the new system is successfully developed (or customised, if it is a vendor product). Structure is developed involving people who implement the technology using defined roles and responsibilities.

- *Training type needed and reasons for not receiving training*: End-users require training programmes and support to help them clearly understand how to effectively use collaborative technologies. Quality training may be delivered by different learning methods.

- *Resistance to the changes involved in introducing new IT and increasing awareness through training*: Stand-alone implementation of collaborative technologies will not be able to enhance collaborative working. Resistance to technology is a combination of resistance to the strategic principle of collaborative working and resistance to the adoption of the technology itself.

- *Efficient collaborative-team-working mechanism*: At the core of collaborative technologies is a synergetic process that relies on all team members contributing to and nurturing a positive team-based organisation based on six dimensions of collaborative working.

- *Most important impacts that should be achieved by collaborative technologies*: New roles and responsibilities should empower employees to challenge their work practices and the impact they have on their current organisational structure and processes.

- *Concerns in relation to shared knowledge*: Sharing project information in collaborative technology can make some team members anxious about the ownership, use and possible abuse of the information they contribute. Such anxieties can also extend to intellectual property.
- *Contractual agreement*: Clients and end-user businesses should seek to protect themselves through contracts. Contracts need amendment to be appropriate for projects where collaborative technology is employed.

This chapter has set out to identify the key factors that enable collaboration between large organisations and SMEs and reports on the findings from the qualitative and quantitative data collection and analysis targeting managerial level and end-users of advanced technologies in AEC SMEs.

10.2 Research methodology

The literature review suggested that the direction of the AEC sector is being tied very much to the implementation of more collaborative working, and that the organisational and end-user aspects are very important for the success of collaborative-technologies implementation. This section describes the methodology to research the validity of these aspects.

Research philosophy

Research philosophy refers to the development of knowledge and the nature of that knowledge (Saunders *et al.*, 2003). Understanding research philosophy, therefore, can help in research design. There are three major ways of thinking about research philosophy: epistemology, ontology and axiology. Researchers are found to relate to research philosophies differently. Although the line between positivism and interpretivism is clear at a philosophical level, Easterby-Smith *et al.* (2002) argued that this line might become blurred at a research design level when the researcher needs to understand the real situation from several perspectives. The researcher might decide to combine the way to obtain data using both qualitative and quantitative data to understand the nature of the real world as perceived by those interviewed and/or surveyed, which is more suited for the purpose of this research.

Research approach

In this research, after the research problem had been defined, it was decided to review the current collaborative-technologies implementation and collaborative-working approaches in AEC companies and to explore the factors that may have contributed to the success or failure and the importance of these factors from the perspective of the SMEs. These objectives are in parallel with the arguments of

Raftery *et al.* (1997), Kuruppuarachchi *et al.* (2003) and Liu (2002), in favour of some degree of methodological liberalism in synthesising paradigms where appropriate in AEC management research. Creswell (2003) provides an example of a scenario in which this approach can be situated: where, for instance, the researcher wants to both generalise the findings to a population and develop a detailed view of the meaning of a phenomenon or concept for individuals, the researcher may first explore generally, in a qualitative manner, to learn about what variables to study, and then study those variables quantitatively, with a large sample of individuals. This scenario mirrors this research and shows that the approach being adopted for this research is appropriate. Therefore, the research approach chosen was triangulated, owing to the nature of the problem defined, based on the literature review findings presented.

Research strategies

The strategy for research is chosen according to the characteristics of the problem. According to Yin (2003), there are many ways to conduct research, governed by the relationship between research questions and research strategy. He suggests that research strategy could be defined by three conditions. Condition 3 indicates that possible research strategies are survey and case study, because the study is focused on contemporary events. According to Yin (2003), a research study could use more than one strategy, and each strategy must be suitable in specific conditions. Therefore, this research was designed to use case study and survey.

Time horizons

Research projects may be cross-sectional or longitudinal (Saunders *et al.*, 2003; Sekaran, 2003). The time horizon of the research was chosen to be cross-sectional, as it was not possible to access all of the organisations for a longitudinal study.

Choices

A research project can implement either qualitative or quantitative methods, or a mixed (triangulated) method. Denzin (1970) identified four types of triangulation in social research, namely: (1) data triangulation, (2) investigator triangulation, (3) theory triangulation and (4) methodological triangulation. In this research, a multi-method was followed, triangulating data, theory and methods. Data from large organisations at managerial level were triangulated with data from SMEs at end-user level. Alternative theories related to implementation of collaborative technologies were used while data were examined. Interviews and questionnaire methods were triangulated, aiming to intensify the reliability and validity.

This section has presented the design of the research methodology adopted in this study. The following section presents the design of the research methodology adopted in this study.

10.3 Data collection strategy

There are many data-collection methods for case study and survey strategies. The case study data-collection methods include, but are not limited to, interviews, observations and past records. Survey data collection is done using questionnaires or structured interviews with the intention to generalise from the sample to a population (Babbie, 1995; Creswell, 2003). It is on record that interviews are the most widely used qualitative method in organisational research (King, 1994). In fact, it was not possible to use methods other than interview for this research. Surveys operate on statistical sampling, which is choosing a representative sample from a population (Fellows and Liu, 2003). Collecting data for surveys through questionnaires has strengths and weaknesses (Neuman, 2005). Therefore, the relevant methods of data collection are explained, respectively.

Interview

The interview can have three forms, namely: structured, semi-structured and unstructured. In this research, semi-structured interviews were adopted for data collection, with two main objectives: to learn and understand the interviewee's perspective on the issue and to know whether the interviewee can confirm insights and information the researcher already holds. The interview questions should be arranged in such a way that each question deals with a separate facet of the topic (Gillham, 2000). The findings of the literature review on collaborative technologies in AEC were used during the design of the interview questions. Table 10.1 summarises the considerations for the interviews.

Questionnaire

The resultant findings of interviews identified the main variables that influence the implementation of collaborative technologies, and these were used to inform the questionnaire survey at an end-user level in SMEs. Close-ended questions with Likert scales were employed to make the questionnaire as easy to complete as possible. Table 10.2 summarises considerations for the quantitative phase.

   This section has presented the design of the research methodology adopted in this study. The following section explores, depending on the previous research, the factors that contribute to the success of implementation of collaborative technologies.

10.4 Data-collection methods

The literature reviewed showed that the main problem in the implementation of collaborative technologies is not technical but related to organisational and end-user levels. The main focus of this section is to describe the selected sample, the purpose of the case studies and the method of data analysis. In order to explore the current collaborative-technologies implementation, qualitative and quantitative data are collected.

Table 10.1 Considerations for the qualitative phase

Unit of analysis: organisation	The research begins with a study of large AEC companies that have experience of collaborative-technologies implementation; this helps develop a basic understanding of collaborative-technologies implementation and barriers; the study then focuses on the SMEs' experience of working collaboratively with large companies in environments enabled by IT
Sample location: United Kingdom	*Selection of case study*: The criteria for selecting the cases are: (a) each selected company is engaged in collaborative arrangements; (b) each company is technology friendly, i.e. already using IT to support part of their business functions; (c) SMEs have a strong business focus and area of activity *Selection of interviewees*: Active managers who have been involved in the implementation of collaborative technologies, actively participated in collaborative technologies, experienced the difficulties and barriers and made decisions to overcome them are considered to yield perspective at organisational level and project level
Sampling strategy: snowball	*Why these*: According to Yin (2003), a single case is used if it represents a critical case or a unique case. Multiple cases are incorporated with the aim to generalise the findings. As the focus is on collaboration in AEC projects, it was clear that there would be more than one case study; otherwise it would not be possible to investigate project organisation level or the factors affecting the collaboration between parties. Therefore, a multiple-case-study strategy was decided. Purposeful sampling is adopted, of which snowball sampling is used, defined as a technique for finding research subjects (Hendricks *et al.*, 1992; Vogt, 1999)
Sample size: 14 companies	*How many*: Patton (2002) argued that the total number of cases is decided during the study. After each case, a quick analysis was made in order to outline the major issues mentioned in the study. Comparing and contrasting the case study with previous ones, the research explored whether it would be enough to make generalisation with the amount of data obtained so far. If it was not enough, a new case was selected in order to achieve either a literal or theoretical replication, following the Yin's guideline (2003), until it was decided that the data obtained from two technology providers, four small, four medium and four large cases were rich enough to make generalisation
Number: 14 interviews	*Why that many?*: Saturation, according to Glaser and Strauss (1967), occurs after many rounds of coding where no new categories emerge from the process. Sample size is deemed to be satisfactory only when the key concepts that have been identified from the collected data have reached saturation point – in this research, when no new data emerge
Tactics: semi-structured (45–60 minutes)	*What to get out of it*: A semi-structured format is chosen in order to ensure each interviewee's response is obtained on the same topics, as well as it being possible to ask further questions when necessary. A face-to-face format was chosen for the interview, as all criteria defined by Gillham (2000) to test the appropriateness of a face-to-face interview are fulfilled

Table 10.2 Considerations for the quantitative phase

Sampling for survey: snowball	The sample used in the survey was drawn from the two interviews with the technology providers. Each was asked to nominate experienced respondents who were users of their technology; only SMEs were selected. It was decided to get a representative from the many specialised professions with useful contribution to projects
Response rate: 53%	Of the 120 questionnaires despatched to the selected sample, 64 were returned; as such, a response rate of 53% was achieved. The AEC sector in the UK is notorious for poor response to questionnaire surveys; 20–30% is believed to be the norm (Black *et al.*, 2000; Takim *et al.*, 2004). Even lower response rates in the region of 14.7% and 11.6% have been described as the norm (Soetanto al., 2002)

Qualitative data

One of the main aims of this study is to explore the factors that contribute to the effectiveness of collaborative technologies as a tool to promote collaborative working in AEC. To achieve this aim, the study looks into insights at the organisational level from a managerial perspective on the list of factors derived from the literature review, by gathering information on:

- the technologies utilised for collaboration, the implementation procedures and their efficiency in relation to collaborative working between large companies and SMEs in the AEC sector;
- the extent to which collaborative-technologies implementations undertaken so far have been successful, and their success criteria;
- thoughts and experience regarding the factors affecting the success of implementing collaborative technologies;
- barriers to and difficulties in collaborative working and the implementation of collaborative technologies.

The semi-structured interview is designed with a mixture of open-ended and closed questions being asked and consists of four sections, namely: (1) profile of interviewees, (2) collaborative-technologies implementation, (3) success level and success criteria of collaborative-technologies implementation, (4) factors affecting the success of collaborative-technologies implementation, and (5) barriers. During the arrangements for interviews, extra attention was paid to ensure that the interviewees:

- were already engaged in collaborative arrangements;
- were technology friendly – that is, were already using IT to support collaboration – and had a strong business focus and area of activity;
- were active managers involved in the implementation of collaborative technologies, actively participated in collaborative technologies, experienced the difficulties and barriers, and made decisions to overcome them.

The qualitative findings from each company are analysed using a combination of qualitative coding and interpretation. Coding is used to organise the raw textual data, which is highly unstructured, as the interviewees referred to the same factors in various questions. Coding is the process of identifying one or more discrete passages of text or other data items that cover the same theoretical or descriptive idea (Gibbs, 2002). Coding is usually a mixture of data reduction and data complication, as it is used to break up and segment the data into simpler, general categories and is used to expand and tease out the data in order to formulate new questions and levels of interpretation (Coffey and Atkinson, 1996). Following the coding principles, the textual data in each transcript were broken into main subject categories by the researcher. The aim was to capture common characteristics and to explore possible relationships, which formed a basis for the interpretations. Interpretation captures the essence of the lessons learned (Lincoln and Guba, 1985). These lessons can be the researcher's personal interpretation or a meaning derived from a comparison of the findings with literature review findings (Creswell, 2003). Therefore, the analysis of the interview data was completed by the interpretation of the results.

Quantitative data

The qualitative analysis revealed partial findings for the main objective of the research, to explore the factors that contribute to the implementation of collaborative technologies in SMEs and to assess their level of importance. One of the main aims of this study was to assess their level of importance by applying quantitative measures. To achieve this, a survey was undertaken targeted at end-users in SMEs who used collaborative technologies to play a role in collaboration. The aim of the survey was to gather information on:

- thoughts and experiences regarding the transformation during collaborative-technologies implementation;
- the role of IT for end-users to promote collaboration in SMEs and their perception of the factors that enable the utilisation of IT;
- end-users' profile influence on the utilisation of IT;
- underlying factors that SMEs should consider when implementing collaborative technologies.

The insights gained from the managerial perspective results stimulated the data collected from a larger number of SME respondents for an end-user perspective. To achieve this aim, the questionnaire consisted of the following six sections, namely: (1) profile, (2) utilisation of technology, (3) organisational environment, (5) sociocultural environment, (6) contractual requirements and economical consequences, and (7) barriers and problems.

Rather than targeting non-collaborative-technology users who may not have relevant experience of collaborative-technologies use only experienced collaborative-technologies users were selected from SMEs.

To aid this investigation, statistical tests were carried out, where appropriate, namely: (a) analysis of attributes, (b) analysis of responses, (c) the chi-squared (χ^2) one sample, (d) Kruskal–Wallis H test, and (e) factor analysis. The rest of this chapter provides a summary of the results generated from theses enquiries.

10.5 Data analysis

By completing the literature review it was possible to derive a list of the prominent collaborative-technology variables that affect the success of implementation. The objective of this section is to ascertain the extent to which companies, within the research sample that used collaborative technologies, considered the variables and how effective they were in facilitating collaboration. To achieve this objective, this section brings together data obtained from the semi-structured interview questions in fourteen case studies. The qualitative analysis revealed partial findings for the main objective of the research, to explore the factors that contribute to the implementation of collaborative technologies in SMEs and to assess their level of importance. The rest of the section assesses their level of importance by applying quantitative measures.

Qualitative data analysis

Brief information on the companies and the interviewees in the fourteen case studies used for semi-structured face-to-face interviews, with mixed managerial levels in AEC companies, are given in Table 10.3.

The technology-providing companies were chosen in order to obtain the perception of the technology, as their responses are based on the implementation experience in various companies. Although availability and accessibility were the main reasons for selection, all fourteen companies were very good targets for the interviews, as they had all been involved in many projects where various types of collaborative technology were implemented.

The literature revealed a number of factors that are said to prevent effective collaboration. The interviewees were asked the extent to which they think it is the case for each factor. The conclusions drawn from the analysis of the case studies are summarised below:

- The ability to motivate people rather than directly guide them is up to the *top-level-management commitment*. Getting users to use the system should not involve coercion, but may involve *vendor support, in-house technical expertise* and *users' technical knowledge and skills*, based on the type of IT, to aid collaboration.
- Top-level management must make it very clear that the new system has to be used in the organisation. They will have two roles: first, to act as collaboration chiefs and manage *electronic information exchange, online relationships* and *online project workflow*, and, second, to ensure that the employees in the organisation *use the features* as agreed by all parties at the beginning of the project.
- The importance of top-level management stems from its role in providing

Table 10.3 Summary of the interviewees

Company	Interviewee's job	Background information
A: Technology provider	Director	Specialist consultancy and analysis in the field of construction collaboration technologies
B: Technology provider	Sales director	Provides highly secure online collaboration software for the AEC sector to manage internal and external collaboration
C: Small	Managing director	Engineering and environmental consultancy, providing support services
D: Small	Architect associate	Dynamic practice set up to deliver architecture
E: Small	Managing director	Provides clients with bespoke management services in a more personal way
F: Small	Design engineer – manager	Offers a diverse range of systems capable of solving complex foundation problems
G: Medium	Director	Urbanists in the widest sense, fusing urban design and landscape architecture
H: Medium	Document controller	Offers a range of development management, project management and construction services
I: Medium	Director	Consulting in engineering design; principle disciplines include civil, structural, geo-technical and geo-environmental engineering
J: Medium	Director	Offers a network of support to projects with a group of specialist architects and technologists
K: Large	Senior design and build coordinator	Group organised into capital works, regeneration and support-services business streams
L: Large	Project leader	Specialises in delivering ambitious and innovative construction projects
M: Large	Site agent	Provider of infrastructure services specialising in the design and delivery of complex engineering projects
N: Large	Architect associate	Interdisciplinary practice of architects, designers, engineers and urbanists

guidance on *conducting online project collaboration, maintaining a non-adversarial environment, aligning collaborative technologies and individual company quality procedures, achieving continuous improvement* and *identifying and resolving unforeseen issues from use of the collaborative technologies* by *defining responsibilities and flow charts of processes*.

• Large companies seek to gain competitive advantage from early implementation by choosing a proactive strategy manifest in process capability, focusing on more customisation of the collaborative technologies to suit core and support processes with greater implementation costs. The relatively high percentage of failure to provide the full benefits expected from the collaborative technologies could be attributed to incorrect or inadequate *inclusion of a new design/redesign of work processes*.

• SMEs acting as external users appear to have less influence where collaborative technologies are usually introduced, because of operational requirements manifest in process performance focusing on management and business network processes. The relatively high percentage of failure to provide the full benefits expected from the collaborative technologies could be attributed to the failure to *inform team members of the common objectives of the project* and *team members' participation with management in defining goals and tasks and creating schedules*.

- Developing self-learning skills such as *online training* and *learning yourself* seem to be viewed more favourably by managers than use of a *hired instructor* or *in-house instructor*.
- The technology provider awareness of the users' background has an impact on *proper training to use the collaborative environments*. The type of IT used can influence the *reasons for not receiving training*. This can be used as an indicator of the *technical support*. The use of *common conventions* will increase the consistency in the collaborative technology, which increases efficiency.

Whereas the interviews explored the factors at an organisational and project level from a managerial perspective, the following section explores the derived factors at an individual level, from end-user perspectives, so that their level of importance can be assessed.

Quantitative data analysis

The insights gained from the managerial perspective results in Abuelmaatti and Ahmed (2014) stimulated the data collected from a larger number of SME respondents for an end-user perspective, as explained in the methodology. The quantitative study of a survey returned by sixty-four respondents was guided by the list of elements proposed by the review developed from the literature. The results suggested that the success of collaborative technologies depends on twenty-three variables (Abuelmaatti and Ahmed, 2014). The main aim of this section is to assess the level of importance of the variables that contribute to the implementation of collaborative technologies in SMEs. Factor analysis is a technique that allows a large number of variables to be distilled into a small number of related factors that could explain most variables that generate the phenomenon under study (Hedderson, 1991; Norusis, 1997; Pallant, 2001). As such, the benefit of this concept for this research is that it reduces many variables into a few factors that best explain the implementation of collaborative technologies in SMEs. In this section, the variables from the questionnaire are analysed. Thus, the factors can be categorised as the representative of the groups of objects that influenced collaborative-technologies implementation in SMEs.

The three main factor-analysis steps are: factor method selection, rational method selection and factor interpretation. The *first step* is focused on a method of extracting factors using either a common factor analysis or a principal component analysis. According to Hair (1998), selection depends upon the objectives of the researcher. If the objective is to minimise the number of variables into underlying factors for predictive purposes, a suitable method should be a principal component analysis. One of the fundamental objectives in this research is to group the objects into factors that will help to describe the essential factors that influence the implementation of collaborative technologies in SMEs. Therefore, principal component analysis was used in this analysis. The factor analysis technique is related to the correlation between each variable. Hedderson (1991) suggested that a correlation matrix of variables should be constructed as a preliminary test. Table 10.4

Table 10.4 Factor loading for the rotated factors

	Component						
	1	2	3	4	5	6	7
Confidentiality concerns	-.883						
Empowerment	.793		-.434				
IPR	.726			.543			
Trust	-.689	-.440					
Contractual agreement	-.683		.511				
Site security	.655						
Training provisions	.634				-.407		
Technical support	.625		.417				
Knowledge creation		.892					
Best practice		.877					
Organisation's income		.863					
Business process		.802					
Relation between users and IT people		.512			.410		
Not singling out for blame (resolving problems in an inclusive manner)			.901				
Sufficient channel of communication			.837				
Reliability and validity of knowledge shared				.733			
Individualism rather than cooperative work (best rewards)				.687			
Flow charts of processes				.616			
Cultural difference				-.543			.458
Non-confrontational working attitudes					.721		
Strategy					-.704		
Informal relationships	.529				.679		
Controlling and coordination by management		.420			.556		
Use of common convention						.821	
Buy-in by all parties						.776	
Management commitment						.755	
Expert status and importance							.882
Defining responsibilities							.734

Notes: Extraction method: principal component analysis; rotation method: varimax with Kaiser normalisation; rotation converged in ten iterations. IPR = intellectual property rights

excludes any variable that has correlation values with the other variables below 0.3 (Norusis, 1997) or 0.4 (Hedderson, 1991) in absolute terms.

The *second step* is to define the numbers of factors. It has also been argued in Field (2000) that, under certain circumstances, the sample size may not be critical. In fact, if a factor has four or more loadings greater than 0.6, then it is reliable, regardless of sample size. It may be argued, therefore, that, where all communalities are greater than 0.6, samples fewer than 100 may be perfectly adequate. These communalities represent the relation between the variables and all other variables. Communalities obtained in this research are shown in Table 10.5, demonstrating 0.537 as the one lower than 0.6 and the lowest obtained.

To further test the adequacy of the data for the factor analysis, two measures – namely, the Kaiser–Meyer–Olkin measure of sampling adequacy and the Bartlett test of sphericity – were obtained (see Table 10.6). According to UCLA (2006), these two tests provide the minimum standard that should be passed. The Kaiser–Meyer–Olkin test varies between 0 and 1, with .50 suggested as minimum in Hair (1998) and Field (2000), and .60 in others. With the Bartlett test, lower

Table 10.5 Communalities

	Initial	*Extraction*
Empowerment	1.000	.926
Confidentiality	1.000	.867
Contractual agreement	1.000	.899
Trust	1.000	.870
IPR	1.000	.903
Informal relations	1.000	.820
Security	1.000	.642
Training provision	1.000	.780
Technical support	1.000	.811
Knowledge creation	1.000	.907
Reuse of best practice	1.000	.904
Increased income	1.000	.885
Business process	1.000	.936
Relationship between users and IT/IS people	1.000	.537
Not singling out for blame	1.000	.888
Sufficient channel of communication	1.000	.802
Attitudes	1.000	.696
Strategy	1.000	.718
Management coordination	1.000	.730
Individualism	1.000	.646
Reliability	1.000	.836
Flow chart of processes	1.000	.784
Cultural difference	1.000	.862
Use of common convention	1.000	.826
Management commitment	1.000	.821
Buy-in	1.000	.817
Expert status	1.000	.837
Responsibilities	1.000	.739

Notes: Extraction method: principal component analysis. IPR = intellectual property rights

than 0.05 significance is required. From the output presented in Table 10.6, it can be seen that, on both counts, the data are suitable for factor analysis.

The analysis saw seven components initially extracted, accounting for 81.027 per cent of the total variance in the twenty-eight variables of IT diffusion at the implementation phase in SMEs, shown in Table 10.7. This extraction of the seven components, according to Field (2000), is based on the Kaiser criterion, which specifies the extraction of all factors with eigenvalues ≥ 1.

Table 10.6 Kaiser–Meyer–Olkin and Bartlett tests of the suitability of the data for factor analysis

Kaiser–Meyer–Olkin measure of sampling adequacy		.571
Bartlett's test of sphericity	Sig.	.000

Table 10.7 Total variance explained by extracted factors

Component	Initial eigenvalues			Rotation sums of squared loadings		
	Total	% of variance	Cumulative %	Total	% of variance	Cumulative %
1	6.699	23.926	23.926	5.220	18.641	18.641
2	5.004	17.871	41.797	4.289	15.318	33.959
3	3.425	12.233	54.030	2.934	10.479	44.438
4	2.479	8.854	62.884	2.685	9.588	54.026
5	2.207	7.883	70.767	2.659	9.496	63.522
6	1.742	6.222	76.990	2.457	8.774	72.296
7	1.130	4.037	81.027	2.445	8.732	81.027
8	.936	3.344	84.371			
9	.820	2.928	87.300			
10	.527	1.881	89.181			
11	.451	1.609	90.790			
12	.380	1.358	92.148			
13	.327	1.168	93.317			
14	.305	1.089	94.406			
15	.286	1.022	95.428			
16	.244	.871	96.300			
17	.194	.692	96.992			
18	.177	.632	97.624			
19	.155	.553	98.178			
20	.121	.432	98.610			
21	.088	.313	98.923			
22	.080	.285	99.208			
23	.067	.238	99.446			
24	.057	.205	99.652			
25	.037	.131	99.783			
26	.030	.109	99.891			
27	.018	.064	99.956			
28	.012	.044	100.000			

Extraction Method: Principal Component Analysis

It is important to recognise that an exact quantitative basis for the number of factors to extract does not exist (Hair, 1998). What exists are a number of criteria, outlined in Hair (1998), Field (2000) and Leech (2004), to mention a few, that are in current use and are applied subjectively in research. One of these is the priori criterion, where the researcher already knows the desired number of components, based, for instance, on theory. In this research, the literature review and the results seemed to indicate five key categories of IT implementation, implying the extraction of five components if this criterion is adopted. Another alternative criterion is the percentage of variance criterion, which specifies that, for social science research, selecting a solution that accounts for 60 per cent of the total variance is satisfactory (Hair, 1998). In this research, 60 per cent of the total variance coincides with five components accounting for 63.522 per cent, as shown in Table 10.7. Indeed, the scree plot produced in Figure 10.1 also provides support for a five-component solution. The cut-off point for selecting components on a scree plot is the point of inflexion or change of direction (Field, 2000). Inspection of the scree plot shows that, after the first five components, increases in the eigenvalues decline.

Ultimately, the aim is to achieve the most representative and parsimonious set of components possible (Hair, 1998). Therefore, the five-component solution was accepted, and the analysis was re-run, extracting five components. These five extracted components account for 63.522 per cent of the total variance in the twenty-eight variables of collaborative-technologies implementation in Table 10.4

Scree plot

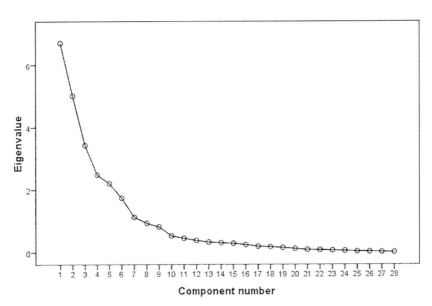

Figure 10.1 Scree plot for component extraction criterion

and satisfy the '7 ± 2' optimum number of dimensions specified by Hofstede and Fink (2007).

The *final step* in the process is the labelling and interpretation of the factors. Principal axis factor analysis with varimax rotation was conducted to assess the underlying structure for the twenty-eight variables of the questionnaire parts mentioned earlier. In order to improve the interpretability of factors, varimax rotation was performed on the extracted component matrix. Varimax rotation is one of a number of rotation techniques. It is recommended as a good approach to simplify the interpretation of factors by maximising the loading of each variable on one of the extracted factors while minimising its loading on all the other factors (Field, 2000; UCLA, 2006). The rotated component loadings show the correlations between each variable, and the component is displayed in Table 10.4, where the items and factor loadings for the rotated factors with loading less than .40 are omitted to improve clarity.

There appear to be five major groupings of factors presented in Table 10.8. The first grouping contains seven individual variables. Second in importance is the grouping of five organisational variables. The third grouping consists of three legal variables. The fourth grouping includes three socio-cultural variables, and the fifth grouping is formed of four project variables.

Cronbach's alpha (α) analysis was conducted to examine the reliability of variables for each factor (Hedderson, 1991; Pallant, 2001). Cronbach's alpha is used to measure how well variables can be constructed into one single factor. Table 10.8 reveals that Factors 2, 3, 4 and 5 fell within the 0.60 or greater range, indicating that these were reliable, and that Factor 1 was *marginally* reliable, as Cronbach's alpha is slightly less than 0.60.

The first factor, which seems to index *individual* orientation, loads more strongly on the first eight items, with loadings in the first column. Three of the items, 'confidentiality concerns', 'trust' and 'contract', indexed low workforce orientation and loaded negatively. This means that confidentiality, established trust and not having to impose contractual agreement would contribute to a positive score on this factor, indicating good relationships within the workforce. The second factor, which seemed to index *organisational* orientation, is composed of the five items with loadings in Column 2 of the table. 'Trust' had its highest loading on the first factor, but had a cross loading on the second factor. 'Controlling and coordination by the management' had its highest loading in Column 5 of the table, but had a cross loading on this factor. The third factor, which seems to index *legal* orientation, comprised the items with loadings in the third column. 'Empowerment' had its highest loading on the first factor, but had a negative cross loading on this factor. 'Contractual agreement' had a negative loading on the first factor, but had a cross loading over .4 on the 'team orientation' factor. 'Technical support' had its highest loading from the first factor, but also had a strong loading from the team orientation factor. The fourth factor, which seems to index *sociocultural* orientation, loads more strongly on the four items in Column 4. Intellectual property rights (IPR) had its highest loading on the first factor, but had a cross loading over .4 on the project orientation factor. The fifth factor, which seems to index the *project* orientation,

Table 10.8 Factor analysis of IT implementation

Factors	Variables	Means
Factor 1: Individual orientation (α = .456, x = 1.8124)	Confidentiality concerns	1.7656
	Empowerment	2.7031
	IPR	1.6094
	Trust	1.7500
	Contractual agreement	1.2812
	Site security	1.5625
	Training provisions	1.6562
	Technical support	1.5312
Factor 2: Organisational orientation (α = .895, x = 2.3062)	Knowledge creation	2.1250
	Best practice	2.0312
	Business incentive	2.5000
	Business process	2.4375
	Relation between users and IT professionals	2.4375
Factor 3: Legal orientation (α = .755, x = 2.026)	Not singling blame	2.3125
	Sufficient channel of communication	2.4844
	Contractual agreement	1.2812
Factor 4: Sociocultural orientation (α = .651, x = 1.7083)	Reliability and validity of knowledge shared	1.9375
	Individualism rather than cooperative work (best rewards)	1.4844
	Flow charts of processes	1.7031
	Cultural difference	1.8438
Factor 5: Project orientation (α = .747, x = 2.7344)	Non-confrontational working attitudes	2.7188
	Strategy	1.5469
	Informal relationships	2.7344
	Controlling and coordination by management	2.7500

Notes: α = Cronbach's alpha; x = Means square

is composed of the items with loadings in the fifth column. 'Relationship between users and IT people' had its highest loading on the first factor, but had a cross loading over .4 on the individual orientation factor.

10.6 Main findings and conclusions

The demands of the AEC sector for collaborative working encourages the development of new and more usable technology to meet these demands. The literature, however, showed that AEC companies adopting emerging IT frequently fail to achieve the full benefits from its implementation. Therefore, this research aimed to identify a set of factors that can enhance collaborative working between large companies and SMEs as key players in the AEC sector, in order to obtain the full benefits from collaborative technologies. This aim was achieved through literature review, case studies and survey results on the implementation of collaborative technologies in the AEC sector. To help develop factors for implementing collaborative technologies effectively within the AEC sector, several approaches were reviewed. The literature, however, showed no theories on collaborative technologies. This led to identification of the indicators of IT readiness in companies and the underpinning theories of IT management. None of these addressed the demands of the AEC sector in relation to the effective implementation of collaborative technologies and the promotion of collaboration in AEC companies. The main conclusion drawn from this review was that the success of collaborative technologies did not only depend on what is introduced to the organisation, but was also related to how it is introduced. The review determined the key elements to focus on during the collaborative-technologies implementation to enhance successful collaboration. The results of this review led to the identification of ten factors.

The findings from the literature review directed the research to employ qualitative and quantitative methods to explore the factors that contribute to the implementation of collaborative technologies. The qualitative method developed a study of fourteen case studies in order to map current practice for the implementation of collaborative technologies and their success level in the AEC sector. All case studies were found to be failing to achieve the full benefits of collaborative-technologies implementation because of underestimation of the organisational, human and legal issues. The case studies revealed a number of factors affecting the success of collaborative-technologies implementation. These factors were affecting collaboration at the organisation level. This also led to further investigation, in the quantitative method, of the factors affecting collaboration at the end-user level. This investigation was guided by the list of factors proposed by the review developed from the literature and suggested that the success of collaborative technologies depends on twenty-three factors, namely: confidentiality concerns, empowerment, IPR, trust, site security, training provision, technical support, knowledge creation, best practice, organisation's income, business process, relation between end-users and IT professionals, not singling blame, sufficient channel of communication, contractual agreement, reliability and validity of knowledge shared, rewards for cooperative work rather than individualism, flow charts of processes, cultural difference, non-confrontational working attitudes, strategy, informal relationships, and controlling and coordination by management.

This research focused on the factors that will enhance collaborative working between large companies and SMEs. The contribution stems from taking a SME's perspective of the problem of failure to achieve the full benefits from collaborative-technologies implementation. Qualitative and quantitative elaboration of collaborative-technologies implementation showed that managerial- and end-user-level analyses are interrelated and provide a holistic approach. The assessment of the indication of factors important to SMEs when adopting IT showed that collaborative technologies' effectiveness does not only depend on what is introduced to the SME, but is also related to how it is introduced. For SMEs to enhance successful collaboration, the study found a list of factors to focus on during the collaborative-technologies implementation to include guidance for implementation. To assess the level of importance of the explored factors, a study of factor analysis was initiated. This allowed the large number of variables to be distilled into a small number of related factors.

References

Abbott, C., Jeong, K. and Allen, S. (2006). 'The economic motivation for innovation in small construction companies', *Construction Innovation*, 6(3): 187–96.

Abuelmaatti, A. and Ahmed, A. (2014). 'Collaborative technologies for small and medium-sized architecture, engineering and construction enterprises: Implementation survey', *ITcon*, 19: 223.

Acar, E., Koçak, I., Sey, Y. and Arditi, D. (2005). 'Use of information and communication technologies by small and medium-sized enterprises (SMEs) in building construction', *Construction Management & Economics*, 23(7): 713–22.

Babbie, E. (1995). *The Practice of Social Research* (7th edn). Belmont, CA: Wadsworth.

Baldwin, N. A. (2004). Overcoming the barriers to the successful introduction of collaborative technologies in construction. In P. Brandon, H. Li, N. Shaffii and Q. Shen (eds), *Designing, Managing and Supporting Construction Projects through Innovation, Proceedings of INCITE 2004*. Langkawi, Malaysia, 18–21 February.

Barker, K. (2012). SME Spotlight: BIM Special [online]. Available at www.bimtaskgroup.org/wp-content/uploads/2012/07/SME-Spotlight_030-033_CN_210612.pdf (accessed 29 July 2012).

Black, C., Akintoye, A. and Fitzgerald, E. (2000). 'An analysis of success factors and benefits of partnering in construction', *International Journal of Project Management*, 18: 423–34.

Cabinet Office. (2011). Government Construction Strategy [online]. Available at: www.cabinetoffice.gov.uk/sites/default/files/resources/Government-Construction-Strategy.pdf (accessed on 22 December 2011).

Coffey, A. and Atkinson, P. (1996). *Making Sense of Qualitative Data*. Thousand Oaks, CA: Sage.

Creswell, J. W. (2003). *Research Design: Qualitative, quantitative, and mixed methods approaches*. Thousand Oaks, CA: Sage.

Denzin, N. K. (1970). *The Research Act in Sociology*. London: Butterworths.

Easterby-Smith, M., Thorpe, R. and Lowe, A. (2002). *Management Research* (2nd edn). Sage Series in Management Research. London: Sage.

e-Business W@tch. (2007). ICT and e-Business in the Construction Industry, Sector Report No. 7. European Commission.

Fellows, R. and Liu, A. (2003). *Research Methods for Construction*. Oxford, UK: Blackwell Science.

Field, A. P. (2000). *Discovering Statistics Using SPSS for Windows: Advanced techniques for the beginner*. London, Thousand Oaks, CA: Sage.

Gibbs, G. R. (2002). *Qualitative Analysis: Explorations with Nvivo*. Maidenhead, UK: McGraw-Hill.

Gillham, B. (2000). *The Research Interview*. London: Continuum.

Gladwell, M. (2001). *The Tipping Point*. London: Abacus.

Glaser, B. and Strauss, A. (1967). *The Discovery of Grounded Theory: Strategies of qualitative research*. London: Wiedenfeld & Nicholson.

Hair, J. F. (1998). *Multivariate Data Analysis*. Upper Saddle River, NJ: Prentice Hall.

Hedderson, J. (1991). *SPSS Made Simple*. Belmont, CA: Wadsworth.

Hendricks, V. M., Blanken, P. and Adriaans, N. (1992). *Snowball Sampling*. Rotterdam, Netherlands: IVO.

Hofstede, G. and Fink, G. (2007). 'Culture: Organisations, personalities and nations', *European Journal of International Management*, 1(1/2): 14–22.

King, N. (1994). The qualitative research inteview. In C. Cassell and G. Symon (eds), *Qualitative Methods in Organizational Research: A practical guide* (pp. 14–36). London, Thousand Oaks, CA: Sage.

Kuruppuarachchi, P. R., Mandal, P. and Smith, R. (2003). 'IT project implementation strategies for effective changes: A critical review', *Logistics Information Management*, 15(2): 126–37.

Leech, N. (2004). *SPSS for Intermediate Statistics: Use and interpretation*. Mahwah, NJ: Lawrence Erlbaum.

Lincoln, Y. S. and Guba, E. G. (1985). *Naturalistic Enquiry*. Thousand Oaks, CA: Sage.

Liu, A. M. (2002). Keys to harmony and harmonic keys. In: R. Fellows and D. E. Seymour (eds), *Perspectives on Culture in Construction*. CIB Report, 275. Rotterdam, Netherlands: CIB.

Manley, K. (2006). 'Frameworks for understanding interactive innovation processes', *International Journal of Entrepreneurship & Innovation*, 4(1): 25–36.

Manley, K. (2008). 'Implementation of innovation by manufacturers subcontracting to construction projects', *Engineering, Construction & Architectural Management*, 15(3): 230–45.

Neuman, W. L. (2005). *Social Research Methods: Qualitative and quantitative approaches* (6th edn). London: Allyn & Bacon.

Norusis, M. J. (1997). *SPSS Guide to Data Analysis*. Upper Saddle River, NJ: Prentice Hall.

Pallant, J. (2001). *SPPS Survival Manual: A step-by-step guide to data analysis using SPSS*. St Leonards, NSW: Allen & Unwin.

Patton, M. Q. (2002). *Qualitative Evaluation Methods*. London: Sage.

Philip, D. (2012). BIM is here and now [online]. Available at www.cnplus.co.uk/hot-topics/sme/bim-is-here-and-now/8632019.article (accessed 29 July 2012).

Raftery, J., Mcgeorge, D. and Walters, M. (1997). 'Breaking up methodological monopolies: A multi-paradigm approach to construction management research', *Construction Management & Economics*, 15(3): 291.

Rezgui, Y. (2011). 'Past, present and future of information and knowledge sharing in the construction industry: Towards semantic service-based e-construction?', *Computer-Aided Design Journal*, 43: 502–15.

Saunders, M., Lewis, P. and Thornhill, A. (2003). *Research Methods for Business Students* (3rd edn). Harlow, UK: Pearson Education.

Sekaran, U. (2003). *Research Methods for Business: A skill-building approach* (4th edn). New York: John Wiley.

Soetanto, R., Proverbs, D. G. and Cooper, P. (2002). 'A tool for assessing contractor performance', *Journal of Construction Procurement*, 8(1): 48–63.

Takim, R., Akintoye, A. and Kelly, J. (2004). Analysis of performance measurement in the construction industry. In S. O. Ogunlana, M. C. Chareonngam, P. Herabet and B. H. W. Hadikusumo (eds), *Globalisation and Construction* (pp. 533–46). Bangkok, Thailand: AIT Conference Centre.

UCLA Academic Technology Services. (2006). Annotated SPSS Output – Factor analysis [online]. Available at www.ats.ucla.edu/stat/spss/output/factor1.htm (accessed 7 December 2015).

Vogt, W. P. (1999). *Dictionary of Statistics and Methodology: A nontechnical guide for the social sciences*. London: Sage.

Whyte, J., Lindkvist, C. and Hassan Ibrahim, N. (2011). Value to Clients through Data Hand-Over: A pilot study: Summary report to Institution of Civil Engineers (ICE) Information Systems (IS) Panel.

Wilkinson, P. (2005). *Construction Collaboration Technologies: The extranet evolution*. London: Spon Press.

Yin, R. K. (2003). *Case Study Research: Design and methods* (3rd edn). London: Sage.

11 Learning from bioclimatic desert architecture

A case study of Ghadames, Libya

Ahmad Taki and Jamal Alabid

This chapter illustrates the use of mixed methodological strategies combining both quantitative and qualitative approaches. These include surveys, interviews, observational techniques, architectural modelling, computer simulation and physical measurements. These methods are all illustrated using a Libyan case study as an example of bioclimatic desert architecture. Energy usage in domestic buildings is responsible for approximately 31 per cent of the total energy consumption in Libya, with the provision of comfortable space conditions forming the major part of this consumption. Contemporary buildings in Ghadames, Libya, adopt air-conditioning systems, which have been recognised as energy-intensive solutions in hot climates. In addition, traditional buildings embrace sustainable features and employ natural ventilation systems to minimise energy consumption. The proposed methodology allows contextualising technical work on thermal comfort, architectural modelling and computer simulation under hot climatic and distinctive sociocultural conditions. It also reviews the results from field surveys undertaken in 2013 and 2014. It shows how 100 residents of Ghadames responded to the environmental conditions and personal well-being, and how such data, together with objective surveys, interviews and observation, can inform architectural modelling, giving a full understanding of the buildings' dynamics, explaining the compactness of traditional houses' urban morphologies, and demonstrating the sociocultural and environmental aspects of the design. These buildings are then tested using computer simulation with EnergyPlus to evaluate the environmental performance and show good agreement with the actual measurements. It was through the use of surveys and interviews that sociocultural attributes could be deciphered and complemented by the physical measurements taken, together developing the new conceptual framework for future housing design.

11.1 Introduction

In Libya, energy usage in buildings is responsible for approximately 60 per cent of total carbon dioxide emissions. Libyan dwellings account for approximately 31 per cent, mainly owing to space cooling for the provision of internal comfort conditions, as shown in Figure 11.1 (GECOL, 2010).

Architecture plays a prominent role in the regulation of indoor conditions, acting as a buffering zone between indoor and outdoor environments. Therefore, architectural passive design solutions should respond to surrounding climatic conditions,

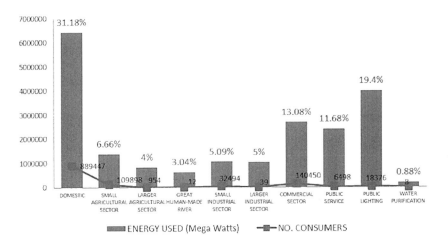

Figure 11.1 Electricity consumption and distribution in Libya
Source: GECOL, 2010

exploiting natural mitigation means to favour users' indoor comfort conditions (Couret *et al.*, 2013). Old, traditional architecture in Ghadames, Libya, shows a unique urban structure that is successful in providing comfortable and liveable residences, combining compactness with minimum exposure to surroundings, without impinging on social privacy (Ealiwa, 2000). The design principles of such architecture, including adaptation to surrounding environmental conditions and the use of natural ventilation, have made this design last for more than 600 years. Furthermore, occupants have adopted ways to deal with such extreme environmental conditions: for instance, the use of different rooms depending on the season. Thus, the harmony and congruence at the level of urban context can be seen clearly with the similarity in form and shape, scale and size, as well as in space organisation and distribution of elements, including materials and colour. These aspects have great psychological impact on residents' satisfaction and on the environmental performance of such a built environment. On the other hand, new, contemporary architecture in Ghadames employs air-conditioning systems that dominate buildings' electricity demand. This suggests that such buildings are not well adapted to local climatic requirements, resulting in significant energy consumption.

The majority of the current studies on Ghadames focus on human thermal sensation and factors influencing people's perception of comfort, in both traditional and contemporary houses (Ealiwa *et al.*, 2001; Elaiab, 2014). Some researchers have studied the environmental strategies in terms of the lighting and ventilation systems of dwellings in Ghadames, and others, the impact of architecture form and relative compactness on building designs as key elements in the response to the outdoor climate (Danielski *et al.*, 2012; McKeen and Fung, 2014). Other research groups focus on the courtyard type of house as the most suitable building form in hot, dry climates (Elwefati, 2007).

The population in Ghadames has increased significantly, resulting in a demand for housing development. These are new, contemporary houses, and, as a result, the problem arises of high electricity demand and a lack of integration of socio-cultural issues. The literature review shows that no other research group has tried to integrate the main sustainable features of old, traditional buildings into new contemporary ones, with consideration to the social integration of the local community and the environmental performance of future building design in hot climates. Therefore, this research focuses on the development of a design framework that would combine the key architectural elements and passive techniques, together with the integration of social needs into the design process and methods to produce a sustainable housing design.

To further address the needs of the people of Ghadames, research must take an integrative methodological approach, one that incorporates subjective measures (e.g. perceptions of comfort and users' behaviour and lifestyle), as well as objective ones (e.g. monitoring, modelling). This study addresses these interrelated issues through the following questions:

1 What factors influence the complex, dynamic feedback between the old, traditional and new, contemporary buildings in the context of Ghadames?
2 How would a multi-method approach allow contextualising technical work on thermal comfort, modelling and architectural designs under very hot climatic and distinctive sociocultural conditions?

This work is timely in addressing the above issues and will produce fundamental understanding, knowledge and practical design guidance for the specific Libyan case context. This multidisciplinary approach also advances knowledge required for improving sustainable housing design. This should help establish a robust methodology that can then be adopted in other hot-climate regions, with different characteristics and variations that would affect the use of building spaces.

11.2 Philosophical stance

It has been prevalently found in the literature that research methodology underpins three main concepts in the research paradigm, as summarised in Sexton (2007): research philosophy, research approach and research techniques. The research philosophy concept in social and science education refers to the two contrasting extremes in philosophy, known as positivism and interpretivism, with a third dimension in between, realism (Almansuri, 2010). Philosophical approaches should lead the researcher to verify the designs appropriate to the problem faced.

11.3 Approach

The way the research is approached often follows the philosophy adopted and depends on the nature of the study. There are a number of issues to consider when choosing one approach over another: for example, the nature of the study and the availability of information on the topic in literature to build a theoretical theme, and, moreover, the timescale and risk of investing in the project and not gaining sufficient data (Saunders *et al.*, 2009). Another point to highlight is that participant preference for an approach often depends on how familiar they are with it. Combining research approaches leads to rigorous and advantageous research outcomes. As such, site investigation and observation in this case were conducted before surveys, to address issues related to the climate, social and architectural prevalence.

11.4 Methodology

The nature of the problem necessitates clear determination of certain tasks, sequence of procedures and proper survey materials and techniques in order to obtain sufficient and useful data on the building and users' characteristics to meet the objectives planned. Introducing design models to new housing development and today's world challenge of sustainable construction requires a profound under-standing of the building context, environment, people and energy sources. Such knowledge can be acquired only through direct observation and experimentation, underlying the correlation between the housing typology, architectural details and the building's environmental and energy performance. Therefore, an explanation of the systematic investigative procedure has been designed to help delineate the path by which the research methods and data-collection techniques would robustly answer the research questions, as shown in Figure 11.2.

This research project adopts a mixed methodological approach combining both qualitative and quantitative methods to provide comprehensive information, resulting in rigorous and reliable research outcomes. Figure 11.2 outlines the methodology and research procedure. It shows how the methods used to collect data form the basis of the design framework and technical advice for future housing design in Libya. This includes a combination of desk and empirical studies. The former involves the analysis of literature and the theoretical basis, and the latter comprises a case study, observations, questionnaires, interviews and full-scale measurements. This combination forms the groundwork for developing an architectural model and the calculation of the entry information for the EnergyPlus simulation, as applied to Ghadames. Figure 11.2 describes how these data have been used to create a design framework demonstrating an understanding of how social integration of the local community is an integral part of the design process.

The proposed methodological approach, as presented in Figure 11.2, was selected as the most suitable research method for this work when 'how' or 'why' questions were posed. The bulk of the data collected is quantitative, responding to the research question, 'How can environmental comfort and architectural design be contextualised under very hot climatic and distinctive cultural conditions, referring to the specific Libyan case context and multidisciplinary approach?'. Therefore, the methods used to collect primary data included a case study and architectural modelling, together with objective and subjective surveys.

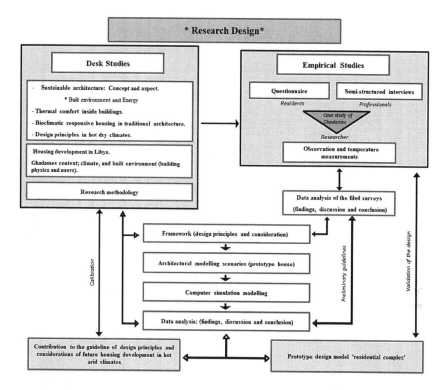

Figure 11.2 Methodology and research approach

11.5 Data-collection strategy

Since 1987, UNESCO has listed the old town of Ghadames as a historic site in the World Heritage List. As such, this makes Ghadames a suitable area to be chosen as a case study shaping the values of old, traditional houses in Libya. It is located in the Libyan Sahara Desert, at an altitude of 350 m above sea level. The climate is characterised by high air temperature, high solar radiation, low rainfall, low relative humidity and many sandstorms. Ghadames provides a unique opportunity for investigation of the differences between old, traditional and new, contemporary architecture. Therefore, it was selected as a case study, and all subjective and object-ive surveys were undertaken in the summers of 2013 and 2014. Prior to the main surveys, a pilot study was carried out in July 2013 as an exploratory stage, in order to determine research problems and test the validity of the proposed methodology. Also, it was an opportunity to gain a fundamental understanding of the nature of the place, referring to residents' culture and social affairs and climate conditions.

Occupants of traditional and contemporary houses were chosen randomly to complete 100 questionnaires, ensuring the majority of the areas of the city were represented. The size of the samples was determined according to recommendations by a number of studies, such as Mccrum-gardner (2010), Ealiwa (2000) and Saunders *et al.* (2009). These questionnaires were translated into local languages. Care was taken to minimise the risk of misunderstandings arising from translations of the words describing the points on various scales by interviewing the respondents and assisting them with completion of the questionnaires. In addition to the subjective study, an objective survey was carried out to measure the four basic environmental parameters as the 100 respondents were completing the questionnaires. Sixteen out of eighteen neighbourhoods were surveyed in the new town, whereas only three out of seven neighbourhoods were visited in the old town, owing to difficulty getting permission from landlords of traditional dwellings. Seven semi-structured interviews with professionals (i.e. educated, administrative, architects and elite groups) were conducted during the visit to traditional dwellings.

Owing to the multicultural backgrounds of the townspeople, descended from three main ethnic tribes – Ghadamesian, Taboo, Tuareg – and an Arab minority, the majority of city residential quarters were surveyed. House location was important in the old town, whereas housing typology was important for selection of the samples of dwellings in the new town. This research employed certain techniques that would enable the data to be validated using a triangulation technique; for example, the semi-structured interviews and personal observation could be used to validate the quantitative data collected by the questionnaires. In addition, desk studies, as shown in Figure 11.2, are a method of triangulation to validate the empirical data collected through field surveys. Studies include the theoretical framework, architectural modelling and computer simulation. This building simulation and case study approach, as well as innovative methods for the collection and calculation of EnergyPlus model inputs for this type of architecture, enable useful, reliable data to be obtained. Such data will be an integral part of the design process for producing sustainable architectural design in hot climates.

11.6　Data collection method(s)

The analysis of published materials helped identify the problem and scope of the study, in order to build a theoretical framework highlighting key elements and issues. In addition to these secondary data, further information was essential to fulfil the objectives, and, therefore, primary data were obtained by both direct observation and subjective surveys, together with architectural and computer modelling. Entry information required for the computer simulation and architectural model would be extracted from design questionnaires, interviews, surveys and physical measurements applied to this case study.

The questionnaires

Questionnaires allow easy collection of a large amount of data representative of the city in question, so that research questions can be precisely answered (Saunders *et al.*, 2009). The questionnaires had several sections: background and personal information; community and personal interaction; thermal environment and personal preferences; occupants' perceptions and comfort; personal well-being; architectural details of dwellings and opinion; together with energy usage and occupant behaviour. The questionnaires were distributed to 100 dwellings: detached and semi-detached houses, villas, flats and other types of dwelling. The sample size was determined considering 10 per cent margin of error, 95 per cent level of confidence and 50 per cent response distribution, according to Mccrum-gardner (2010) and 'Raosoft' sample size calculator and other references. Figure 11.3 shows the design and conduction process of the questionnaire, from pilot surveys, through redesigning, to framework design.

Semi-structured interviews

The qualitative data provided by semi-structured interviews offer validity and reliability. According to Saunders *et al.* (2009), the typology of interviews may vary, depending on the nature of the questions, from highly formalised to informal conversations, whereas semi-structured interviews are often used to understand the relationship between certain variables in explanatory studies. The study adopted

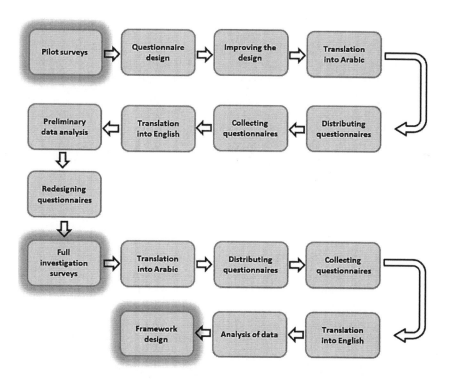

Figure 11.3 Questionnaire procedure

semi-structured interviews owing to their ability to fit with other techniques and because they allow interviewees to ask questions or add comments. According to Healey and Rawlinson (1994), these interviews were non-standardised, one to one and face to face.

Interviews were conducted in four traditional houses and three contemporary houses. The interviews covered the following:

- the concept behind the adaptability of traditional patterns of houses to climate and social life;
- hindrances locals and authorities may encounter when integrating the traditional house concept;
- reasons behind the prevalence of contemporary housing designs;
- the efforts, hindrances and people's awareness of sustainability issues;
- the implications of architectural and technical solutions for energy usage and the environment.

Observational research

A major impediment to accessing houses in both the old and new towns was the reluctance of the society to cooperate in the study. Permission was sought from landlords to visit traditional dwellings well in advance, to allow the tribes to prepare their houses to receive a stranger. Visits to modern houses were limited to those people known by the research group. Additionally, owing to the civil war in Libya, the study was only able to survey buildings in selected areas that were pre-arranged with local authorities. All of these issues led to some technical and logistical problems in the survey process.

Part of this work was implemented concurrently with subject interviews, monitoring and assessment of their thermal conditions. Observations involved recording, describing, analysing and interpreting people's behaviour. This could only be achieved after relationships and trust had been established with tribe leaders (Saunders *et al.*, 2009), where the researcher became a member the community, in order not merely to observe but also to experience. Photographic work was essential to document and assess dwelling conditions, typologies and design elements in both traditional and modern settlements. The main outcomes of the observational research can be summarised as follows:

- direct investigation of respondents' thermal perception and preference;
- observing the building characteristics in both traditional and contemporary dwellings;
- experiencing and feeling the difference between the indoor thermal conditions of traditional and contemporary environments;
- sketches and photographs of settlements to create a demonstrative conceptual model of residential buildings.

Physical measurements

Subjects were asked to complete a questionnaire at the same time as air temperature, globe temperature, surface temperature, air velocity and relative humidity were being recorded. All subjects were sitting on the floor, and sensors to measure the environmental parameters were placed at a height of 0.3 m above the floor, representing the centre of gravity of the subject. Air temperatures were recorded using radiation-shielded thermocouples and were logged every 15 minutes, and average values were calculated every hour. Air velocity was measured using an omnidirectional anemometer. The mean globe temperatures were measured, using a standard globe thermometer, and mean radiant temperatures were then calculated. The equipment used in this study complied with criteria given in ISO7726 standard. Outdoor environmental parameters were also recorded.

Architectural modelling and computer simulation

The data collected from the field surveys, including building users' characteristics and opinions, together with building characteristics and details, were used to calculate the input data required for creating architectural drawings and EnergyPlus input data. The former resulted in the production of a 3D model to visualise a part of a neighbourhood in the old town and a typical contemporary house using Sketchup and 3D max tools. The latter resulted in prediction of the environmental performance of these buildings.

11.7 Data analysis

Architectural modelling

A typical traditional house in Ghadames is approximately 30 m², and the total building area is approximately 80 m². It is a four-storey building, with a guest room on the ground floor. The living room is located on the first floor and opens to a mezzanine floor area where private bedrooms are located. The stairs lead to the kitchen and summer shed space on the upper floor. The house is adjacent to other houses and sheltered on four sides, creating well-shaded environments while allowing natural light and ventilation through the small roof openings.

A 3D architectural model was created showing how the whole urban fabric works as one complex to protect the indoor environment from direct exposure to outdoor extreme climate conditions and its relationship with the surrounding green belt (Figure 11.4).

Analysis of subjective surveys

The first two sections of the questionnaire highlighted the personal and sociocultural aspects of participants. Of the respondents, 93.5 per cent were aged between 20 and 50 years, with 63 per cent male and 37 per cent female. The majority were

Figure 11.4 Demonstration of condominium of twenty-two houses in old town

educated and employed in either higher-education or governmental sectors. The average family size of those participants was 6.0 people, and 39.1 per cent of them have lived in the old town for at least 6 years. Eighty-five per cent of subjects indicated living next door to neighbours and having relied on either relatives or neighbours as sources of support. In most cases, neighbours are relatives. Although the majority of the neighbourhood know each other, 67 per cent of them agreed that the current community lacks social interaction, owing mostly to changes in lifestyle. The majority of respondents prefer the mixed housing designs of traditional and contemporary style, and 63 per cent of them have refurbished their contemporary houses for reasons of privacy. The findings show that the total floor

Table 11.1 Total dwelling area according to housing typology

Total dwelling area (m²)	Villa	Detached house	Semi-detached house	Flat
Mean	241.68	180.38	145.00	125.00
Median	270.00	180.00	125.00	125.00

Table 11.2 Monthly electricity usage according to housing typology

Monthly electricity usage (kWh)	Villa	Detached house	Semi-detached house	Flat
Mean	1,179	1,371	1,232	1,143
Median	1,429	1,358	1,286	1,143

Table 11.3 Pearson correlation coefficient of monthly electricity usage to total dwelling and living-room areas

	Correlations	Living-room area		Correlations	Dwelling area
Energy usage (kWh)	Pearson correlation	0.342	Energy usage (kWh)	Pearson correlation	0.114
	Sig. (2-tailed) N	0.032 39		Sig. (2-tailed) N	0.245 39

area of new dwellings is not an appropriate indicator for estimating the energy consumption. Table 11.1 shows total floor areas for different dwelling types, and Table 11.2 shows that there was no considerable difference in total energy consumption across all housing typologies. This has led to a study of the architectural details of individual houses and cooling patterns to establish the parameters influencing energy consumption.

The survey indicates that 76.1 per cent of respondents use two AC units in bedrooms, and 82.6 per cent of residents use one AC unit in the living-room, which is kept running throughout the day. Table 11.3 tests the subject effect of total dwelling floor area compared with living-room area using the Pearson correlation coefficient. The effective coefficient value for any positive Pearson correlations must be higher than 0.3 and with a p-value of less than 0.05 according to Doane and Seward (2011). Thus, there was a correlation between the area of the living room and total energy consumption represented in 0.342 Pearson coefficients and 0.032 Sig. values, whereas there was no effective correlation subject to the dwelling floor area.

The survey shows that respondents predominantly agree that social-life and climate-related issues are respected in traditional architecture; meanwhile, the majority stated that contemporary houses offer greater aesthetic values and more spacious designs. Nevertheless, it was understood that the urban and individual designs of houses in the new town are not well adapted to climate requirements and are unable to respond to local cultural affairs. They heavily rely on mechanical and artificial air-conditioning systems, resulting in significant energy consumption. The majority preferred housing designs that mixed traditional and contemporary styles, indicating that old settlements do provide a high degree of privacy and respect social environmental affairs, adapting well to local climate requirements, but do not satisfy today's society needs.

Application of ISO7730 (ISO, 2005)

A seven-point ASHRAE sensation scale for investigating thermal comfort and preference was used, where −3 is cold, +3 is hot, and 0 is neutral. A sample of the results of the measurements in old and new houses is illustrated in Table 11.4. It shows residents generally felt neutral to slightly warm in old houses, even when the air temperature was approximately 32°C. It can be seen that there are clear discrepancies between predicted mean votes (PMVs) and actual mean votes (AMVs) for the old houses. Table 11.4 also shows good agreement between PMVs and AMVs of the occupants in new, air-conditioned houses. They relied more on air conditioning being turned on to achieve the indoor air temperature of 25°C.

The results from the field surveys show that female respondents preferred lower air-conditioning-cooling set-point temperature of 18°C, compared with male respondents who preferred 20°C in new houses. This could be owing to their clothing, where women often wear the traditional garments with higher clothing values. In this regard, the preferred cooling set-point temperature is found to have an impact on the amount of energy used by air-conditioning systems. A positive correlation was found between this set-point temperature and space cooling hours.

Computer simulation

Typical models of traditional and contemporary houses were created so that their environmental performance could be assessed using computer simulation with EnergyPlus. Figure 11.5 shows the sun-path diagrams for old and new houses on a typical summer day.

The entry information was calculated for the EnergyPlus input model. This included construction details, activity, openings, layout, orientation, HVAC systems, lighting, heat gains, etc. Simulations were carried out in the summer season – May, June, July and August – to predict energy consumption, temperature profiles and daylight factor distributions. Tables 11.5 and 11.6 show sample results of the indoor temperatures in both old and new houses, respectively. These results are in agreement with the measurements carried out in these houses: see Table 11.4. It can also be seen that the predicted relative humidity level is lower than the actual values, owing to water streams crossing underneath the urban structure of the old town, as well as the green belt of thousands of trees surrounding the urban fabric of the town.

Energy consumption in a new house was also predicted, showing an average monthly electricity usage of 1,570 kWh. The energy consumption was the highest in July, at 2,660 kWh. The actual electricity reading in July for the house was 2,775 kWh. The CO_2 emissions associated with the cooling demand were predicted depending on the type of AC system and its coefficient of performance (COP). The COP was found to be approximately 3.6.

The distribution of daylight factor was also assessed using computer simulation. Figure 11.6 illustrates visualisations of spaces in old houses showing the average daylight factor of approximately 0.5 per cent, which is much less than the recommended value as presented in BS 8206 (BSI, 2008). Old houses use reflective surfaces and mirrors inside living-rooms to redirect the daylight deeper into the space through successive reflections, while keeping the roof void small to maintain privacy.

Table 11.4 Sample of measurements and respondents' thermal perception in old and new houses

Housing type	T_{globe} °C	T_{wet} °C	T_{dry} °C	Rh (%)	V_a (m/s)	AC T_{set} °C	Met★	Clo★	AMV	PMV	PPD(%)
Old (living-room)	32.5	19.16	31.56	30.9	0.06	None	1.2	0.48	0.81	2.21	83.6
New (guest room)	25.0	17.1	25.16	35.4	0.18	16.66	1.1	0.55	-1.0	0.706	16.96

Note: ★ Values were taken from ASHRAE, 2013

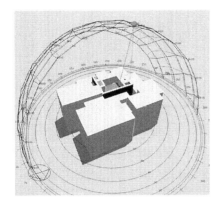 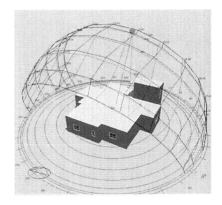

Figure 11.5 (a) Typical old traditional house model for computer simulation. (b) Typical new contemporary house model for computer simulation

Table 11.5 Indoor thermal conditions in an old house with natural ventilation system

	Month			
	May	June	July	August
Air temperature (°C)	31.34	31.77	32.00	32.07
Radiant temperature (°C)	32.84	33.43	33.67	33.73
Operative temperature (°C)	32.09	32.60	32.84	32.90
Outside dry-bulb temperature (°C)	34.19	35.24	36.21	36.40
Relative humidity (%)	18.59	16.20	12.41	14.92

Table 11.6 Indoor thermal conditions in a new house with AC units

	Month				
	May	June	July	August	September
Air temperature (°C)	23.33	23.95	24.32	24.37	23.71
Radiant temperature (°C)	27.43	28.62	29.01	29.07	27.99
Operative temperature (°C)	25.38	26.28	26.67	26.72	25.85
Outside dry-bulb temperature (°C)	32.05	35.24	36.21	36.44	33.04
Relative humidity (%)	26.13	25.36	18.84	23.07	28.73

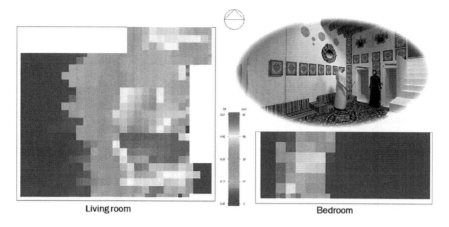

Living room   Bedroom

Figure 11.6 Level of natural lighting in a traditional house

11.8 Main findings and conclusions

This research represents the second phase of an ongoing 3-year research project looking into the development of a new design framework for housing developments in hot climates. The framework evaluates sustainable features of old, traditional buildings, and these can be fed into new, contemporary ones, with consideration to the social integration of the local Ghadames community and the environmental performance of future building design. An extensive amount of case data was collected and is being analysed for the development of this new framework. This chapter showed how a multi-method approach enabled such work to be carried out on thermal comfort and design in hot climates and in distinctive sociocultural conditions. The combination of surveys, interviews, observations and physical measurements and the mixture of qualitative and quantitative data achieved rigorous and reliable results and facilitated the architectural modelling and computer simulation by calculating the entry data to feed into the EnergyPlus simulation program. It was through the use of surveys and interviews that sociocultural attributes could be deciphered, and these complemented the physical measurements taken, together developing the new conceptual framework for future housing design.

Regarding this case, the findings to date can be summarised as follows:

- Residents felt neutral to slightly warm in traditional buildings, even when indoor temperatures reached 32°C. However, in contemporary buildings, occupants felt neutral to slightly cool.
- The architectural model created for a residential condominium in the old town helped to demonstrate the sociocultural and environmental aspects considered in the design of traditional houses.

- Old settlements of Ghadames showed greater response to the climate, employing natural ventilation during the day and night and taking advantage of the very compact urban structure to minimise heat gain and offer great mutual shadowing.
- The majority of participants preferred mixed housing designs of traditional and modern styles, indicating that old settlements do provide a high degree of privacy and respect for social environmental affairs but do not satisfy today's society needs.
- The results from computer simulations are in good agreement with actual measurements from the mixed methodology approach, encouraging its use in the future modelling of conceptual architectural housing designs.

References

Almansuri, A. A. (2010). *Climatic Design as a Tool to Create Comfortable, Energy-Efficient and Environmentally Wise Built Environment*. PhD thesis, University of Salford, UK.

American Society of Heating, Refrigeration and Air Conditioning Engineers (ASHRAE). (2013). *ASHRAE Handbook—Fundamentals*. Atlanta, GA: ASHRAE.

BSI. (2008). *BS 8206–2 Lighting for Buildings – Part 2. Code of practice for daylighting*. London: BSI.

Couret, D. G., Díaz, P. D. R. and de la Rosa, D. F. A. (2013). 'Influence of architectural design on indoor environment in apartment buildings in Havana', *Renewable Energy*, 50: 800–11.

Danielski, I., Fröling M. and Joelsson A. (2012). 'The impact of the shape factor on final energy demand in residential buildings in Nordic climates'. World Renewable Energy Forum, WREF 2012, including World Renewable Energy Congress XII and Colorado Renewable Energy Society (CRES) Annual Conference, pp. 4260–4.

Doane, D. P. and Seward, L. E. (2011). 'Measuring skewness', *Journal of Statistics Education*, 19(2): 1–18.

Ealiwa, M. A. (2000). *Designing for Thermal Comfort in Naturally Ventilated and Air Conditioned Buildings in Summer Season of Ghadames, Libya*. PhD thesis, De Montfort University, Leicester, UK.

Ealiwa, M. A., Taki, A. H., Howarth, A. T. and Seden, M. R. (2001). 'An investigation into thermal comfort in summer seasons of Ghadames, Libya', *Building & Environment*, 36(2): 231–7.

Elaiab, F. M. (2014). *Thermal Comfort Investigation of Multi-storey Residential Buildings in Mediterranean Climate With Reference to Darnah, Libya*. PhD thesis, University of Nottingham, UK.

Elwefati, N. A. (2007). *Bio-Climate Architecture in Libya: Case studies from three climatic regions*. MSc dissertation, Middle East Technical University, Ankara, Turkey.

General Electric Company of Libya (GECOL). (2010). Annual Report 2010 [online]. Available at www.gecol.ly (accessed 25 May 2015).

Healey, M. J. and Rawlinson, M. B. (1994). Interviewing techniques in business and management research. In V. J. Wass and P. E. Wells (eds), *Principles and Practice in Business and Management Research* (pp.123–46). Aldershot, UK: Dartmouth.

International Organisation for Standardisation (ISO). (2005). *ISO 7730:2005 Ergonomics of the Thermal Environment – Analytical determination and interpretation of thermal comfort using calculation of the PMV and PPD indices and local thermal comfort critieria*. Geneva, Switzerland: ISO.

Mccrum-gardner, E. (2010). 'Sample size and power calculations made simple', *International Journal of Therapy & Rehabilitation; Research Methodolgy Series*, 17(1): 10–14.

McKeen, P. and Fung, A. (2014). 'The effect of building aspect ratio on energy efficiency: A case study for multi-unit residential buildings in Canada', *Buildings*, 4: 336–54.

Saunders, M., Lewis, P. and Thornhill, A. (2009). *Research Methods for Business Students* (5th edn., p. 649). Harlow, UK: Pearson Education.

Sexton, M. (2007). *Axiological purposes, ontological cages and epistemological keys*. PhD thesis, Research Institute for the Built and Human Environment, University of Salford, UK.

Part V

Action research

12 Embedding action research in the built environment

An action approach

Lloyd Scott

The aim of the chapter is to provide the opportunity to consider the potential for theorising, in the specific area of built environment education, as to what action research can contribute to a researcher's armoury. It will clarify the role of action research and its various linkages between theories, models, hypothesis generation and the different types of empirical work. The development of the ability to evaluate and select research methods appropriate to the investigation of research questions in the built environment will be explored through a case study example. The chapter will address the following: what action research is; the purpose of conducting action research; the philosophical worldview of action research, with particular reference to the built environment; the development of action research; and the models and key characteristics of action research. It will also discuss the role of the researcher in action research and ethical considerations. The chapter will include a worked case study set in the built environment and entitled, 'Research: Embedding formative assessment-led learning strategies [FALLS] in construction management education'. The issue for any researcher in regard to the choice of appropriate research method is addressed by exploring the qualitative nature of the research question. The reader will be prompted to reflect, a cornerstone principle of an action research approach, on the philosophical position of the researcher, the choice of a qualitative approach and the methods used to gather the qualitative data.

12.1 Introduction

Human beings have always shown an interest in, and concern to come to terms with, their environment and try to make sense of and understand the nature of the phenomena their senses experience (Cohen *et al.*, 2007). In the educational environment, research exists in many forms and can range across a variety of topics and engage in a diverse array of methodologies and methods to answer particular research questions. All research, including built environment (BE) educational research, needs to be subjected to careful and considered methodological assessment and rigour. Traditional forms of theory are grounded in the logic of binary divides (McNiff and Whitehead, 2005) and thus can tend to exclude the notion of practitioner research. For example, this might be an academic applying other people's theories to their practice, or, as is the case here, a researcher engaged in examining other people's practices and extracting a theory or framework. The primary concern, in relation to methodological issues, is to ensure that the research is conducted within the parameters of a methodology that appropriately addresses its aims. Therefore, an approach that accommodates change, such as an action methodology or participatory action research (AR) methodology, seems to offer an appropriate framework within which to explore and understand the assessment practices of BE academics on undergraduate higher-education programmes.

This chapter includes a discussion around the modes of enquiry engaged with in the context of this research enquiry. The research itself is a study of the views of students and the researcher in the field of construction management on assessment, followed by an exploration of the 'FALLS' practice that was developed to help to lead to an improvement in how students learn in construction education.

AR – which is also often referred to as participatory action research (PAR), community-based study, cooperative enquiry, action science and action learning – is an approach commonly used for improving conditions and practices in a range of environments (Whitehead, 2003). It can be used to involve practitioners conducting systematic enquiries in order to help them improve their own practices, which in turn can offer the potential to enhance their working environment and the working environments of those who are part of it – students, practitioners (contractors), clients and the like. The purpose of undertaking AR is to bring about change in specific contexts, as Parkin (2009) describes it.

Although qualitative approaches to research can cause the novice researcher some problems, the case study will suggest a strategy and method for reaching the decisions around that approach. The exploratory nature of AR as an appropriate research process will be explained and justified, and its relevance as a 'reflective practice' and a 'practice-based approach' will be made clear. The underpinning of the assumption that, in research, 'thinking' and 'doing' are continuously interrelated and mediated by decisions will be reinforced for the reader by their being prompted to contribute and engage in problem scenarios offered in the chapter. Through their observations and communications with other people, action researchers are continually making informal evaluations and judgements about what it is they do. This approach can offer an embedded approach to reflection and so lead to improvement.

12.2 Philosophical positioning of action research in the built environment

One of the critical decisions that any researcher will need to make when designing their research study/project is on the paradigm within which they will situate their work. The term 'paradigm', which is derived from the work of the famous science historian Kuhn, refers to a set of general philosophical assumptions about the nature of the world (ontology) and how one can understand it (epistemology). These are often assumptions that tend to be shared by those researchers working in a specific field or tradition. Paradigms will also typically include specific methodological strategies linked to those assumptions. At the more abstract and general levels, examples of such paradigms are philosophical positions that include positivism, constructivism, realism and pragmatism. Each of these embodies very different ideas about reality and how we can gain knowledge of it. Although well beyond the scope of this chapter and book to describe these paradigms and how they can inform a research study, it is imperative that any good research study/project and the researcher associated with it understand the fundamentals in terms of the philosophical position underlying the research project.

Many action researchers position themselves within the 'pragmatic' paradigm. As research is about the generation of knowledge, so the purpose of AR is to learn through action, which may then lead on to personal or professional development. The essence of the type of enquiry conducted by an action researcher is that it involves an investigation of some component or aspect within a social context, where the creation of impressions and the transmission of information are required. Philosophically, the researcher who chooses to embed their work in an AR framework is required to select and make a decision about what methodology to adopt, while also reporting on findings. The researcher will need to consider their ontological and epistemological stance. Whichever philosophical stance is taken, it is important to declare this and show an understanding of the implications of doing so with regard to data collection and analysis. In order to do that, a closer look at what the different theoretical perspectives mean within the context of AR is necessary.

Ontology, as mentioned above, is used to designate the theory of being. Its mandate is the development of strategies that can illuminate the components of people's social reality – about what exists, what it looks like, the parts that make it up, and how these parts interact with each other (Blaikie, 1993: 6). Within AR, researchers attempt to consider this reality as socially constructed and not external or independent. This meaningful construction happens through interpretations of researchers' experiences and communication. The research journey they present will be based on subjective accounts from the characters within their research environment. The methods of data collection that are used will be consistent with their ontological stance.

The term 'epistemology' presents a view and justification for what can be regarded as knowledge – what can be known and the criteria that knowledge must satisfy in order to be called knowledge rather than beliefs (Blaikie, 1993: 7). For traditional researchers, knowledge is certain and can be discovered through scientific means. For an action researcher, the nature of knowledge and what constitutes knowledge are different. The type of data collected is more subjective where the experience and insights are of a unique and personal nature. What people say and how this might be interpreted are important for an action researcher, especially when they relate to knowledge creation. Thus, it is imperative, in any reporting of their research and claims about knowledge generation, that action researchers acknowledge their epistemological stance and make a defence of that position. An important, linked matter is that of understanding the implications of doing so with regard to data collection and analysis.

AR is more of a holistic approach to problem-solving, rather than a single method for collecting and analysing data. Thus, it allows for several different research

Table 12.1 Attributes that separate action research from other types of research

Aspect of research approach	Approach and reasoning
The philosophy (epistemology) of the research and why	Usually a pragmatic approach, as AR is about change
Research approach and why	To bring about development in practice by analysing existing practice and identifying elements for change
Description of the strategy and reason for the chosen strategy	Research on action by using action as a tool for research
Research choices or methodology and why	Mixed methods or multi-methods (both qualitative and quantitative)
Time horizons	Can be cross-sectional, as AR is best at focusing on change over time
Methods (techniques) used for data collection	Observation, field notes, questionnaires, reflection, interviews (structured or semi-structured)

tools to be used as the project is conducted. Table 12.1 provides a table of the several attributes that separate AR from other types of research. Primary is its focus on turning the people involved into researchers, too – people learn best, and more willingly apply what they have learned, when they do it themselves. It also has a social dimension – the research takes place in real-world situations.

12.3 Research methodology

AR comprises a family of research methodologies that aim to pursue action and research outcomes at the same time. It therefore has some components that resemble consultancy or change agency, and some that resemble field research. The focus is action to improve a situation, and the research is the conscious effort, as part of the process, to formulate public knowledge that adds to theories of action that promote or inhibit learning in behavioural systems. One of the key characteristics of this approach is collaboration, which enables mutual understanding and consensus, democratic decision-making and common action (Oja and Smulyan, 1989: 12).

In the academic literature, discourse on AR tends to fall into two distinctive camps. The British tradition, which is especially linked to education practice, tends to see AR as research oriented towards the enhancement of direct practice. Carr and Kemmis, for example, refer to it as 'simply a form of self-reflective enquiry undertaken by participants in social situations in order to improve the rationality and justice of their own practices, their understanding of these practices, and the situations in which the practices are carried out' (1986: 162).

In this sense, the action researcher is a practitioner, an interventionist seeking to help improve processes of engagement. From Argyris *et al.*'s position, it 'takes the form of creating conditions in the behavioural world of the client system that are conducive to inquiry and learning' (Argyris *et al.*, 1985: 137). It can offer a situation where lasting improvement and change involve the participatory action researcher assisting participants to change themselves so that their interactions will create these conditions for enquiry and learning. Hence, to the aim of contributing to the practical improvement of problem situations and to the aim of developing community knowledge is added a third aim of AR, to develop the self-help competencies of people facing problems. Within this broad definition, there are four basic themes:

1   collaboration through participation;
2   acquisition of knowledge;
3   social change; and
4   empowerment of participants.

The process that the researcher uses to guide those involved can be seen as a spiral of AR cycles, consisting of phases of planning, acting, observing and reflecting (Masters, 1995). As Oja and Smulyan (1989) point out, the underlying assumption of this approach – which can be traced back to Lewin's writing in 1948 – is that effective social change depends on the commitment and understanding of those involved in the change process (p. 14). In other words, if people work together on a common problem, 'clarifying and negotiating ideas and concerns', they will be more likely to change their minds if research indicates such change is necessary. Also, it is suggested that collaboration can provide a community of people with the time and support necessary to make fundamental changes in their practice that endure beyond the research process (Oja and Smulyan, 1989: 14–15).

So, the role of the action researcher is somewhat similar to that proposed for contemporary facilitators in helping communities identify and adopt more sustainable natural-resource management practices. These facilitators may come from the community or they may be research or agency staff. However, their most effective role will be to involve the wider community in developing participatory attitudes, excitement and commitment to work together on jointly negotiated courses of action to bring about improvements and innovation for individual and community benefit. Although this role is similar to much of consultancy, AR provides a means that is more rigorous and that allows for the development of public knowledge to advance the field.

By establishing and developing conditions for the development of others, the action researcher acquires increasing skills in such things as the ability to build shared vision, to bring to the surface and challenge prevailing mental models, and to foster more systematic patterns of thinking. To paraphrase Senge *et al.* (1994), action researchers are responsible for building frameworks and networks through which people are continuously expanding their capabilities to shape their future. That is, action researchers are responsible for developing a learning environment that challenges the status quo and for generating liberating alternatives (Argyris *et al.*, 1985: xi). Accordingly, the general aims of AR are frequently expressed in terms of orienting process criteria (e.g. participation, emancipation), and it seems worthwhile to continue to stress these characteristics to differentiate AR from other approaches to social change (Altrichter *et al.*, 2013).

The AR case study that is being presented in this chapter involved cycles of interactions between the researcher and the research subjects that had two separate objectives:

1   participants learning to build capabilities and enhance ability to understand and generate constructive behavioural change through reflective action on assessment practice;
2   generation of research findings for the researcher on the usefulness of FALLS as a tool to enhance such actions by the participants to improve their learning.

Iterative cycles of learning and development of research understandings, however, can be challenging to balance, as the motivations and benefits of the different parties can compete with each other (Steinfort, 2010). In this case study, the AR process described by Brannick and Coghlan was adopted because the research objectives aligned with their broad definitions of AR should be (Brannick and Coghlan, 2010: 4):

- research in action, rather than research about action;
- a collaborative democratic partnership;
- research concurrent with action; and
- a sequence of events and an approach to problem-solving.

The high level of participation of the case study subjects in constructing an understanding of their own environment represented relevant learning for them, and their reflections on the application and impact of these techniques in their own environment formed a rigorous foundation of the research evidence (along with supplementary data from surveys and interviews).

Another reason an AR approach was chosen was because it combines both learning and generation of new research knowledge through a series of iterative cycles across multiple cases. This was considered to be an ideal way to explore the research objectives in such an emergent situation. The learning aspect of the research is that the participants learn some techniques that they then try to implement as part of their own environment, and they then reflect on what resulted from that (and ideally go through iterative cycles to further develop their capabilities on the use of the methods to continue to enhance their ability to influence others and enhance outcomes). The research aspect is to understand how the application of the FALLS impacted the participants (Scott and Fortune, 2011, 2013). In these situations, it was not possible to predict what the outcomes would be, and the participants needed to use reflection before and after the intended action in order to optimise the outcomes and determine reasons as to why they got the results they did. The role of the researcher in this research was to guide the participants in the use of the FALLS, assist with planning the intended interventions and collate reflections from the participants. An aligned perspective was presented from McKay and Marshall (2001), who proposed the 'dual imperatives of action research' as being problem-solving (a form of learning within a specific issue context) and a contribution to research. They explained the difficulties of researching the process being taken while actually applying the process to create outcomes in terms of developing solutions to real problems, and they suggested the AR approach was ideal for this.

The action research process in identifying the problem

Identifying a topic for AR is one of the most important steps in the process:

*   AR topics should address realistic classroom problems or issues.
*   Research topics should also be weighed against several practical considerations, including your personal interest in the topic, its potential importance, the amount of time it will require, the anticipated difficulty, potential costs and any ethical issues.
*   Narrowing a topic can be accomplished by addressing practical considerations and also through self-reflective, descriptive and explanatory activities. Preliminary information related to the topic should be gathered.
*   This information can be gathered by talking with other educators, reviewing curricular materials or examining professional publications. This helps establish a connection between your given project and what has been done before.

Literature reviews can provide guidance in helping to identify and narrow a topic, formulate research questions and hypotheses, select appropriate data-collection methods and identify appropriate techniques for data analysis. When you are reviewing related literature, it is important to consider its quality, objectivity and timeliness. When you are trying to locate related literature, it is best to begin with secondary sources and then move to primary sources. Furthermore, it is best to focus your review on primary sources. If it becomes necessary to write a formal review of related literature, bear in mind its purpose: to convey to all individuals interested in the topic the historical context of the topic, the trends experienced by the topic, how theory has informed practice and vice versa. A documented literature review should not consist of an annotated list of summaries of research, but rather it should flow smoothly for the reader as a cohesive essay.

12.4 Data-collection strategy

The identification of the research problem for the case study was not difficult. The previous construction technology and applications module, with two end-of-module summative and time-constrained examinations, did not encourage students to engage in the type of 'learning by understanding' advocated throughout more recent 'constructivist' educational literature. Subsequently, the use of rote learning was very apparent. Also, the pressure placed on the lecturer to deliver large amounts of curricular content for examination purposes was leading to an environment led by traditional lecture-centric teaching methods and to surface learning on the part of the students. The majority of the participants showed great enthusiasm to learn about their discipline but most found it difficult to learn under the pressure of the curriculum content and associated summative examinations. Research evidence suggests that to increase the engagement and activity levels of students the use of assessment strategies such as project work, group work and problem-based learning is recommended (Rust *et al.*, 2005).

Preparing BE students for real-life work should involve engaging them in tasks where they must make complex judgements about their own work and prepare them for making decisions in the uncertain and unpredictable circumstances in which they will find themselves in the future (Boud and Falchikov, 2006). This should involve moving assessment from the domain of the assessors into the hands of the learners, making it an indispensable accompaniment to lifelong learning (Boud, 2000).

Students engaged with the FALLS research also got the chance to work collaboratively with their colleagues and participate in peer learning. The potential benefits of peer learning have long been recognised, but many existing assessment practices act to undermine these and lead students to reject learning cooperatively (Boud and Associates, 2010). The use of group projects and collaborative learning also encourages students to learn from other students, as well as from the lecturer (Keppell *et al.*, 2006).

Participants

Participants in the study included the author, other lecturing colleagues and two classes of construction management students (one from the second year of the undergraduate programme and the other from the third year of that programme). In the case of these groups, there was no necessity to gain parental consent, as all participants were adults. Data collection was carried out by the author as the researcher, which provided valuable insights. Other lecturers (who taught on parallel modules) were peripherally involved during the first phase and became active participants during the second phase, particularly in regard to the impact of the changes made in the module included in the research project.

Qualitative data gathered and analysed from questionnaires, focus-group interviews, observational and reflective diaries culminated in findings to show that the FALLS learning paradigm significantly improved the competence, understanding, motivation and confidence of those students participating in the research. Noticeable improvements in other key skills, such as group participation, reflective learning and self-assessment, also emerged through this pedagogical implementation. The data-analysis section presents the collective qualitative findings from the data collated (questionnaires, observations, focus groups and interviews) throughout the three cyclical phases of the research project. As this was an AR project, the analysis of qualitative data continued throughout the life of the project and was, therefore, not a self-contained phase of its own.

Throughout the research, the attitude of the students towards the FALLS approach was, without exception, extremely positive. Among the contributing factors to this was a greater awareness of their responsibility for their own learning, where the freedom to choose, plan and carry out the assessment-oriented tasks provided this. There appeared to be genuine willingness and interest in learning among the students, where they saw a clear relationship to the subject area and a means of applying this knowledge to real-life practical situations. The highest motivational aspect to learning that emerged from the findings came from the authenticity of the projects and their relevance to real-world design work and problems. This was evidenced by the quality of both the draft submissions and the final projects presented by the candidates.

Ethical considerations in action research

Research involving humans is subject to considerations of ethical conduct, norms that guide the relationship between researchers and participants, with a view to protecting the latter from suffering harm, disrespect or unfair treatment. Informed by leading international ethics norms, the AR researcher must recognise that the search for knowledge aimed at understanding and improving our world is a fundamental human endeavour. This implies that research has its own requirements, those of academic freedom, to be balanced against three competing sets of ethical considerations: respect for persons, concern for welfare and justice. Chevalier and Buckles (2013) refer to the Canadian Tri-Council in their research and they proffer three principles around ethics in AR that are quite appropriate, particularly in reference to research practices that are critical, action-oriented and community-based.

The first principle recognises the inherent worth and dignity of all human beings. Researchers are under an obligation to respect the autonomy and freedom of individuals and groups to deliberate about a decision and act on it. Respect for people implies that researchers must seek the free, informed and ongoing consent of all those participating in research, with authorised third-party involvement in the case of persons lacking the capacity to decide. This is particularly important in situations of power imbalance and potential coercion or undue influence affecting the relationship between researchers and participants. Consent about participation must be informed: this means that measures must be taken to ensure that participants have a reasonably complete understanding of the purpose of the research, what is involved and any risks and potential benefits that may result from participation, at least those that can be foreseen. It is good practice to ensure that this voluntary, informed consent is obtained through a signed consent form; this is usually a requirement of the ethics process in university research regulations.

The second principle involves the obligation to protect the welfare of living individuals or groups by not exposing them to any unfavourable balance of benefits and risks associated with participation in the research, especially those that are serious and probable. As privacy is a factor that contributes to people's welfare, confidentiality tends to be the norm. Respect for privacy regulates the control of personal information about participants and those who are important to them – information that is identifiable and not in the public domain. Confidentiality is usually obtained through the collection and use of data that are anonymous (e.g. survey data) or anonymised (irrevocably stripped of direct identifiers).

The final principle refers to justice, where there is a requirement that all people be treated with equal respect and concern for fairness and equity. Criteria of appropriate inclusion are needed to ensure that no particular people or groups (defined by age, gender, language, religion, ethnicity or disability) bear an unfair share of the direct burdens of participating in research. Nor should any population or group be overprotected or discriminated such as to be excluded from research without appropriate justification, arbitrarily depriving them of the potential benefits of participation (e.g. information-sharing, capacity-building, community action learning and problem-solving). Concern for justice also calls for mechanisms to identify the dual or multiple roles of researchers and those assisting them, with a view to disclosing, minimising or eliminating related conflicts of interest, whether real, potential or perceived.

While complying with the institutional requirements for ethics in research, the case study presented adopted the three principles advocated by Chevalier and Buckles (2013).

12.5 Data-collection methods

This research study was carried out in just one subject from a very large curriculum in which rote learning is seen to be extensive. With very few exceptions, the students who participated in this study agreed that they found the FALLS method of learning, teaching and assessment more productive for their learning needs, more engaging through challenging, real-life design work, and more resourceful in providing them with the competencies they need to work within their chosen discipline. What is generally agreed throughout the research literature is that modern society requires a fundamentally different conceptual discourse for assessment (Clegg and Bryan, 2006). Such assessment activities should not only address the immediate needs of certification of students on their current learning, but also contribute in some way to their prospective learning (Boud and Falchikov, 2006).

Society now demands more than passive graduates who have complied with a rigid regime, and employers and professional groups are placing expectations on institutions to deliver graduates who are prepared for, and can cope with, the real world of work (Boud and Falchikov, 2006). Student-centred learning can foster knowledgeable, competent, reflective and committed learners (Mentkowski and Associates, 2006) who are better prepared for the unorthodox types of real-work problem that are typically associated with engineering and BE disciplines. Students may escape from poor teaching through their own activities, but they are trapped by the consequences of poor assessment, as it is something they are required to endure if they want to graduate (Boud and Falchikov, 2006). The more we can engage students in assessment activities that are meaningful to them and that contribute to their learning, the more satisfying will be their educational experience (Boud and Falchikov, 2006).

Data collection took place over the period February to May 2012 for the research-investigation part of this enquiry. During that period and beyond, the data collected were assembled and analysed in a phased approach. Each phase of the research process was documented, and the data were recorded appropriately, as per good practice. What emerged from the focus-group sessions was that most participants reported that they had never used formative assessment methods in any previous class setting. In the early stages of the research, the reasons for this were not evident, as most of the participating group suggested their level of mathematical ability was between average and good. However, further analysis of the data showed that, although their understanding levels were sufficient to learn construction detailing for summative assessments, the relationship and link between these calculations and real-world construction-method tasks were unclear to most students.

Student E, in a focus-group session, commented that, 'I always had the ability to do calculations in my exams, but I didn't know why I was doing them'. There appeared to be general agreement with this from the group. Student C also reported that, 'I didn't know where in construction or on site these calculations could be used, they were just written on the board and we were told to follow them for the exam'. From the students' perspective, there seemed little reason to continue with any craft calculations or mathematics when they could see no practical value to do so. Based on the evidence of the projects submitted by students, along with the research data analysed, it appears reasonable to suggest that the majority of the students have the ability to learn construction-detailing calculations and are happy to use them, once they can see a clear, practical purpose to do so. There was a general consensus among research participants that, having gone through the FALLS method applied to calculations in construction, they could now use this knowledge and apply it to a real building service design. Student F commented that, 'I could now see what the construction details were doing in the project tasks; it's more like what is expected of us in industry'.

There was general consensus among each group that learning through FALLS was a better learning experience than they had had previously. Many students expressed a sense of enjoyment in the learning experience and thought that other subjects should be delivered in this manner. During the focus-group interviews, it was most interesting to listen to students still discussing knowledgeably with each other the mathematical formulas they had used in the various projects a number of weeks previously. When questioned on this, they expressed an ease of understanding, having used these formulas to produce an end product. Student A made the point that, 'when I did construction calculations before I finished with a number that meant nothing. This time I had to go and find certain aspects to match and also make sure it would fit in the space. Now it means something to me'. Most of this group agreed that they now had a deeper understanding of this subject and they had also learned where the calculations could be used within construction detailing.

From the research data, there is strong evidence to suggest that the students strongly engaged and were more motivated to broaden their learning about this subject area. When given an opportunity to learn and apply it to their chosen field, they willingly undertook the tasks and were interested in coming up with a workable solution. Most of the comments relating to this in the focus-group interviews and classroom discussions concurred that studying questions and subjects for an end-of-term summative assessment was not increasing their competence, knowledge or skills in the chosen area of study. In fact, all of the focus-group participants agreed with the words of Student F: 'it would have been great to do all our subjects and design them within a project building. I would definitely have learned much more about construction management'. In addition, Student C followed on to say: 'if we were assessed on this it would be much better and we would learn far more than trying to remember questions for an exam'. Student B also made the point that:

> It would be great to see how all the different systems like project planning and on-site supervision join up with each other in a building project. At the moment we just do everything separately for examinations. It is just read out and I can never see how they all link together.

It would appear from the responses that the students indicated complete satisfaction with the FALLS approach, but also, more particularly, with the linking of the assessment to their learning. The use and practice of applied processes that are task-oriented, with feedback, contributed to a distinct increase in (construction) confidence levels over the course of the research study. With this increase in confidence, many students, throughout the research period, showed initiatives to extend their (construction) skills beyond the curriculum. An example of this was the extremely high attendance during and after class throughout the semester. It is the contention of the researchers that students attend class if there is a reason to do so.

One of the major aims of this research project was to provide a platform that enabled students to gain a deeper understanding of construction technology and its application in the construction process, through a task-oriented approach, with a formative-assessment focus. The findings of each individual AR cycle hold that the majority of those who participated in the research reported a distinct increase in both their level of understanding and their competence in the use of applied construction methods.

The early interviews with students indicated that their confidence around the use and understanding of the application of calculations was lacking. However, later findings revealed that they gained confidence in their knowledge and understanding through the task-oriented approach. Student B commented:

> The tasks we did helped me get a better understanding of why and where the construction details could be used to building design. I now also feel more confident in my ability to use them in relation to my job.

Most felt they would now be more able to accomplish many design-oriented tasks and solve work-related problems from the experience of this approach. The feedback from the majority of students, together with all of the research findings, has shown that a task-oriented approach, with the underpinning of a strong formative-assessment strategy, when used in the correct circumstances, has the capacity to create learning environments that are facilitative to deep levels of learning and understanding.

The findings and discussions presented in detail above, along with positive feedback from students during the course of this study, leave us confident to report that we believe a deeper understanding and application of applied calculations can be achieved through the implementation of FALLS and assessment within a construction detailing module.

12.6 Data analysis

'Analysis is the interplay between researchers and data. It is both science and art' (Strauss and Corbin, 1998: 1). 'Data analysis is a way of "seeing and then seeing again". It is the process of bringing order, structure, and meaning to the data, to discover what is underneath the surface of the class-room' (Hubbard and Power, 1993: 65). It is one of the most difficult and yet crucial aspects of qualitative research (Basit, 2003). The collection of the above data through questionnaires, observation and diaries means very little until it is analysed and evaluated (Bell, 2005). It is important to acknowledge that, although there are structured approaches proffered in the literature, each individual researcher will have their own way of making sense of the data, and their analytical approach, although supported by the depth of research analysis literature, will be what works best for them. It is, therefore, important for the novice researcher to 'dip their toe in the water' and grapple with that analysis. In this case study, as I sought to make sense of the data I collected, I interacted with it, questioning what had happened. I sought to delve through the layers of the classroom, uncovering relationships that had much deeper meaning than I initially observed. To uncover these relationships, I engaged in indexing, writing memos and employing the constant comparative method. Although 'few people write about their processes as they unfold' (Hubbard and Power, 1993: 67), it is important to be reflective and capture the essence of each interaction as soon as possible after the event. The use of a reflective diary, whether electronic or otherwise, is a huge must for the action researcher.

What is described are some of the processes that helped me make sense of the data collected to address the research questions. In organising the data collected, I indexed according to unit of study. Within the unit of study, I subdivided the data into background information, classroom environment, curricular structure, student talk, teacher talk, book talk and others. This indexing, and the creation of a table of contents, made it easier to progress in the analysis. Such a process also helped identify the most important data, which in this case were the conversations that took place in large and small groups, supplemented by field notes. As the data were read and reread, sections that were representative of certain categories were pinpointed and, following that, grouped by index. This process also helped to narrow the focus of the study to using previous research to discuss the interrelationship between student engagements in formative-assessment practices. In indexing, what was constructed was a framework inductively taken from the fieldwork (Schatzman and Strauss, 1973). Also, sense was made of the data, as memos were written from early in the study, 'theorizing write-up about codes and their relationships as they strike the analyst while coding' (Glaser, 1978: 43). What is reported is the findings of this analytical process that comprise the last step in the constant comparative method – summarising and describing the findings.

Category formation represents the heart of qualitative data analysis (Creswell, 2009). As this is a sequential, mixed methods research enquiry, the analysis of qualitative data took place as each phase of the research was completed. To facilitate data analysis and research-project management, a computer-assisted qualitative data analysis software (CAQDAS) package, NVivo 9, was used. Use of such software is increasingly common in qualitative research (Bringer *et al.*, 2004). Creswell (2009) suggests that CAQDAS provides an effective system for storing, locating and accessing large amounts of data easily. Furthermore, Bringer *et al.* (2004: 251–2) argue that CAQDAS allows complex data searches, affords opportunities to improve data security, and 'offer[s] the "revolutionary" prospect of demonstrating methodological congruence because of a level of transparency that is so labour intensive that it is rarely, if ever, seen in manual methods'. Furthermore, Coffey *et al.* (1996: 172) recommend that, 'anyone now embarking on a sustained piece of qualitative research should seriously consider the potential value of computer-aided storage and retrieval'. It was used to categorise and code qualitative interviews, as it was not considered feasible to do this by means of solely analysing the transcripts. Therefore, the qualitative data collected for this study have been electronically categorised, segmented, coded and summarised for inclusion in the findings.

> Coding has a crucial role in the analysis of these data and above all it allows the researcher to communicate and connect with the data to facilitate the understanding of the emerging phenomena and to generate theory grounded in the data.
>
> (Basit, 2003: 152)

In selecting the research methods, the aim was to choose those that were suitable for this particular case study and were readily analysed, interpreted and presented (Bell, 2005).

For this medium-scale research project, it was useful to use Miles and Huberman's view of qualitative data analysis (1994). They define qualitative data analysis as consisting of three concurrent flows of activity:

1   *Data reduction*: the process of selecting, focusing, simplifying, abstracting and transforming the data that appear in written-up notes or transcriptions.
2   *Data display*: an organised compressed assembly of information that permits conclusion-drawing and action.

3   *Conclusion-drawing and verification*: the meanings emerging from the data that have to be tested for their plausibility, their sturdiness and their conformability, which is, in effect, their validity (Miles and Huberman, 1994).

These three types of flow of analysis activity and the data-collection activity itself can form an interactive cyclical process, as shown in Figure 12.1. This is particularly suited to the cyclical nature of a sequential, mixed methods approach to research. The researcher steadily moves between these four 'nodes' to make up the general domain called 'analysis' for the whole of the study (Miles and Huberman, 1994).

Some further direction on the analysis of the interview phases, the online survey data and results was taken from Krueger (1998), who also considers data analysis as a fluid process rather than a series of isolated tasks. Reinforcing this view, he makes the point that although some steps occur simultaneously there is a need to loop back and repeat an earlier step (Krueger, 1998), as Figure 12.1 demonstrates. Basit (2003) also supports this view, emphasising that data analysis is not a discrete procedure carried out in the final stages of a research project, but is an all-inclusive activity that continues throughout the life of the project.

Assigning numerical values to text data can be a very useful tool, either as part of a larger project that employs many different methods or as a basis for a complete piece of work, as is the case for this research. With the use of sophisticated software packages such as SPSS, it is relatively easy to deal with the computation side of things, and it is possible to come up with numerous tables and charts almost instantly, once the data are installed. However, it is important that the underlying principles

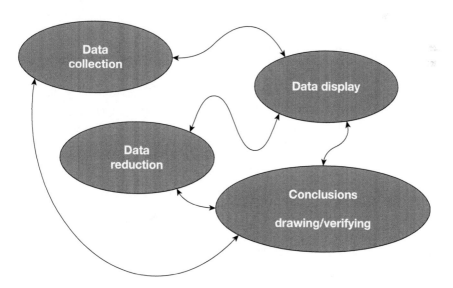

Figure 12.1 Data-analysis model used for this research enquiry

of statistical analysis are both understood and employed, if sense is to be made of the results generated by such quantitative software. The quantitative analysis of the online survey was completed through the software package SPSS, providing the option to do some correlations and analysis.

During the course of this research, awareness and reflection were maintained of and on the impact the research environment might have, on account of the closeness of the researcher to the material. With regard to the attempt to ensure that this does not affect the validity of the project, Wilson (2004) states that the researcher must address the work with 'reflexivity'. This is to report what the researcher's impact is and, in cases where it is contrary to the essence of the research, to eliminate it (Wilson, 2004). It was, therefore, the intention of this researcher to avoid subjectivity, be reflexive and assess the effects that this might be having on the process and include these findings in the discussions.

12.7 Reflections

The struggle for congruency between our theories and practices is a commonality among action researchers. Torbert proffers it clearly: our practice 'aims toward greater congruity between the values one espouses and the values one enacts' (2001: 111). The author recalls this struggle in the early formative stages of the implementation of participatory AR, including a realisation that one's approaches to research and evaluation may have been incongruent with the values of the empowering, non-formal approaches to formative-assessment practices we espoused in our work outside academia.

The potential contributions of AR to social change are limited if we are a marginal force within universities, and yet the challenges of scaling up, a measure of the acceptance of AR in the development arena, are equally daunting. Weaknesses of AR include its localism and the difficulty we find in intervening in large-scale social-change efforts, especially in the higher-education setting. The bulk of AR takes place on a case-by-case basis, often making huge advances in a very local situation, but then extending beyond that local context. The strength of AR is that impact can be made at a local level, and, although there are some who consider that research such as AR should not be given the same position as larger, more quantitative approaches, AR has an important place in research.

In regard to the case study presented, many of those students who participated in the research still make contact (mostly via email, IM and text messages) about the impact of this assessment approach on their learning. They reflect and report on their memories of our classes and how they allowed them to see and develop a more diverse approach to their learning, but, most of all, it provided them with an opportunity to explore and achieve the best in themselves. My sharing this research enquiry and the findings is in no way intended to convey a naive, heroic or triumphal tone. It is, nevertheless, most rewarding to share the reflections about small change and the transformation that has impacted on a small number of students. The author remains hopeful that that experience shaped and changed their lives for the better.

It is hoped that, through the sharing, description and analysis of this very concrete case study example, learners, educators and researchers will come to recognise students' intelligences and cultures in the learning environment through dialogues initiated from multicultural literature, while critically exploring hard issues, such as diversity, gender equity, strength and power, and considering multiple perspectives. I do not offer an answer, as there is no single one; what I do offer are perspectives: my perspective and that of the students who endeavoured to problematise their learning issues around assessment as a way to enhance their learning.

As an educator, I have learned much through AR alongside my students, while merely seeking to provide access for all learners to a more structured and beneficial approach to formative assessment. I would like to offer encouragement to all academics and those in education development to begin 'to re-create and rewrite [these] ideas' (Freire, 1970: xi) and to start implementing learner-centred, responsive pedagogies through deep explorations and dialogues. I would further encourage the use of the FALLS in the classroom as a strategy for challenging the perceived issues that surround assessment practice.

Finally, AR projects test knowledge in action, and those who do the testing are the interested parties for whom a base result is a personal problem. AR meets the test of action, something generally not true of other forms of social research. Conventional researchers worry about objectivity, distance and controls. Action researchers worry about relevance, social change and validity, tested in action by the most at-risk stakeholders.

In this chapter, we have tried to give the reader an overview of what is entailed in carrying out AR and the purposes of carrying out AR projects. The presentation of frameworks and definitions of AR can only give a hint of the flavour of the experience – to digest the nature of AR fully you need to be an active participant. Expert views, from those who have contributed to the development and a more widespread acceptance of AR, have been indicated, and their names and publications have been cited as landmarks in the progress of the methodology. A salient feature of AR is its cyclical structure, and this can be highlighted by the use of diagrammatic forms. Different readers will, indeed, react to each diagram differently and use them as they see fit within their own action plans.

The key characteristics of the AR approach have been explored. Some key theoretical underpinnings, associated with AR, have been briefly presented. The example of a previously conducted AR case study project has been used as the vehicle to provide the context for using AR and provided to enable the reader to become familiar with the various processes and stages prior to experiencing them personally.

References

Altrichter, H., Feldman, A., Posch, P. and Somekh, B. (2013). *Teachers Investigate Their Work: An introduction to action research across the professions.* Abingdon, UK: Routledge.

Argyris, C., Putnam, R. and Smith, D. (1985). *Action Science: Concepts, methods, and skills for research and intervention.* San Francisco, CA: Jossey-Bass.

Basit, T. N. (2003). 'Manual or electronic? The role of coding in qualitative data analysis', *Educational Research*, 45(2): 143–54.

Bell, J. (2005). *Doing Your Research Project: A guide to first-time researchers in education, health and social science* (4th edn). Maidenhead, UK: Open University Press.

Blaikie, N. (1993). *Approaches to Social Inquiry*. London: Polity.

Boud, D. (2000). 'Sustainable assessment: Rethinking assessment for the learning society', *Studies in Continuing Education*, 22(2): 151–67.

Boud, D. and Falchikov, N. (2006). 'Aligning assessment with long-term learning', *Assessment & Evaluation in Higher Education*, 31(4): 399–413.

Boud, D. and Associates (2010). *Assessment 2020: Seven propositions for assessment reform in higher education*. Sydney, NSW: Australian Learning and Teaching Council.

Brannick, T. and Coghlan, D. (2010). *Doing Action Research in Your Own Organization*. London: Sage.

Bringer, J. D., Johnston, L. H. and Brackenridge, C. H. (2004). 'Maximising transparency in a doctoral thesis: The complexities of writing about the use of NVivo within a grounded theory study', *Qualitative Research*, 4(2): 247–65.

Carr, W. and Kemmis, S. (1986). *Becoming Critical. Education, knowledge and action research*. Lewes, UK: Falmer.

Chevalier, J. and Buckles, D. (2013). *Participatory Action Research: Theory and methods for engaged inquiry*. Abingdon, UK: Routledge.

Clegg, K. and Bryan, C. (2006). 20 reflections, rationales and realities. In C. Bryan and K. Clegg (eds), *Innovative Assessment in Higher Education* (pp. 216–27). Abingdon, UK: Routledge.

Coffey, A., Holbrook, B. and Atkinson, P. (1996). 'Qualitative data analysis: Technologies and representations', *Sociological Research Online*, 1(1) [online]. Available at www.socres online.org.uk/1/1/4.html (accessed 3 December 2015).

Cohen, L., Manion, L. and Morrison, K. (2007). *Research Methods in Education* (6th edn). London: Routledge Falmer.

Creswell, J. W. (2009). *Research Design: Qualitative, quantitative, and mixed methods approaches*. Thousand Oaks, CA: Sage.

Freire, P. (1970). *Pedagogy of the Oppressed*. New York: Continuum.

Glaser, B. (1978). *Doing Grounded Theory: Issues and discussion*. Mill Valley, CA: Sociology Press.

Hubbard, R. S. and Power, B. M. (1993). *The Art of Classroom Inquiry. A handbook for teacher–researchers*. Portsmouth, NH: Heinemann.

Keppell, M., Au, E., Ma, A. and Chan, C. (2006). 'Peer learning and learning-oriented assessment in technology-enhanced environments', *Assessment & Evaluation in Higher Education*, 31(4): 453–64.

Krueger, R. A. (1998). *Analysing and Reporting Focus Group Results* (vol. 6). London: Sage.

McKay, J. and Marshall, P. (2001). 'The dual imperatives of action research', *Information Technology & People*, 14(1): 46–59.

McNiff, J. and Whitehead, J. (2005). *All You Need To Know About Action Research*. London: Sage.

Masters, J. (1995) The History of Action Research. In I. Hughes (ed.), *Action Research Electronic Reader*, The University of Sydney [online]. Available at: www.behs.cchs.usyd. edu.au/arow/Reader/rmasters.htm (accessed 22 February 2016).

Mentkowski, M. and Associates (2006). *Learning that Lasts: Integrating learning, development, and performance in college and beyond*. San Francisco, CA: Jossey-Bass.

Miles, M. and Huberman, A. M. (1994). *Qualitative Data Analysis: An expanded sourcebook* (2nd edn). London and Thousand Oaks, CA: Sage.

Oja, S. N. and Smulyan, L. (1989). *Collaborative Action Research: A developmental approach.* New York: Falmer Press.

Parkin, P. (2009). *Managing Change in Health Care Using Action Research.* London: Sage.

Rust, C., O'Donovan, B. and Price, M. (2005). 'A social constructivist assessment process model: How the research literature shows us this could be best practice', *Assessment & Evaluation in Higher Education,* 30(3): 231–40.

Schatzman, L. and Strauss, A. L. (1973). *Field Research.* Englewood Cliffs, NJ: Prentice-Hall.

Scott, L. and Fortune, C. (2011). 'Formative assessment practices in built environment higher education programmes and the enhancement of the student learning experience', *Association of Researcher in Construction Management Proceedings,* Bristol, UK, 2011.

Scott, L. and Fortune, C. (2013). 'Towards the improvement of the student experience of assessment and feedback in construction management education', *European Journal of Engineering Education,* 38(6): 661–70.

Senge, P. M., Kleiner, A., Roberts, C., Ross, R. and Smith, B. (1994). *The Fifth Discipline Fieldbook: Strategies and tools for building a learning organization.* New York: Doubleday.

Strauss, A. and Corbin, J. (1998). *Basic of Qualitative Research: Grounded theory procedures and techniques.* Newbury Park, CA: Sage.

Torbert, W. R. (2001). The practice of action inquiry. In P. Reason and H. Bradbury (eds), *The Handbook of Action Research: Participative inquiry and practice* (pp. 250–60). London: Sage.

Whitehead, J. (2003). 'How are the living educational theories of master and doctor educators contributing to the education of individuals and their social formations?' Paper presented at BERA Annual Conference, Heriot-Watt University, Edinburgh, 12 September 2003.

Wilson, T. P. (2004). 'Qualitative "versus" quantitative methods in social research', *Bulletin de Méthodologie Sociologique,* 10(1): 25–51.

13 Information systems to support planning and decision-making in construction organizations using action research methodology

Salman Azhar

Action research is an established research method for scholarly investigations in built environment, information systems, project management, and social and medical sciences. It is an iterative approach in which the researcher investigates the problem domain, diagnoses the problem, gets involved in introducing some changes to improve the situation, and evaluates the effects of those changes (Naoum, 2001). Action research aims at building and testing theory within the context of solving an immediate practical problem in a real setting. It combines theory and practice, researchers and practitioners, and intervention and reflection. The method produces highly reliable research results because it is grounded in practical action, aimed at solving a realistic problem situation while carefully informing theory. This chapter illustrates the use of the action research approach by presenting a case study of the development and implementation of information systems that support planning and decision-making in construction firms. Construction organizations typically deal with large volumes of project data containing valuable information. It is found that most organizations do not use these data effectively for planning and decision-making purposes. Based on the action research philosophy, a framework is developed to guide the storage, processing, and retrieval of validated and integrated data for timely decision-making and to enable construction organizations to redesign their IT infrastructure and organizational structure matched with information-system capabilities. A functional model and a prototype system were developed using the data-warehousing technique to test the framework. The results revealed significant improvements in data management and decision-support operations that were examined through various quantitative (time-savings and operational cost savings) and qualitative measures (ease of data access, data quality, response time, productivity improvement etc.). The case study demonstrates that action research provides an effective platform to increase collaboration between academic researchers and construction industry practitioners by focusing on the solution of real-life construction problems. It simultaneously assists in practical problem-solving and expands scientific knowledge, as well as enhancing the competencies of the researchers and practitioners.

13.1 Introduction

Academic research in applied disciplines such as built environment, information systems (ISs), project management, and social and medical sciences has the dual mission of generating theoretical and conceptual knowledge and simultaneously contributing to the solution of practical problems (Azhar *et al.*, 2010). Research in such disciplines has often been criticized for focusing on theoretical and conceptual issues and neglecting the needs of the industry. Thus, academic research loses relevance and has little or no impact on practice (Azhar, 2007). Close collaboration between researchers and industry practitioners can ensure that the research results would be acceptable and applicable in the industry.

A novel approach to improve collaboration between researchers and practitioners in academic research is provided by the action research method. This qualitative research method is unique in the way it associates research and practice through change and reflection (Rezgui, 2007). Action research aims to build and test theory within the context of solving an immediate practical problem in a real setting. It combines theory and practice, researchers and practitioners, and intervention and reflection. The method produces highly reliable research results because it is grounded in practical action, aimed at solving a realistic problem situation while carefully informing theory (Baskerville, 1999).

Action research is also known as a problem-solving approach. In action research, the researcher reviews the existing situation, identifies the problem(s), gets involved in introducing some changes to improve the situation and evaluates the effect of those changes. This type of research is more attractive to researchers, practitioners and students from a professional background who have identified a problem during the course of their work and wish to investigate and propose a change to improve the situation (Naoum, 2001). To understand the action research method, let's consider the following problem situation.

The research problem

The project data in construction organizations are typically stored in operational and applications databases of ISs to support construction operations and decisions. The data are often non-validated and non-integrated and stored in a format that makes it difficult for the decision-makers to make quick decisions. This is a typical problem in many organizations and it is due to the data-modeling limitations of existing ISs, traditional organizational setups and inadequate information technology (IT) infrastructure (Ang and Teo, 2000; Azhar and Ahmad, 2007). Moreover, in construction organizations, the databases are likely to be geographically and/or functionally dispersed, thereby making them difficult and inconvenient to access in the relatively short time available for decision-making. A decision-maker may have to wait days or weeks for responses from subordinates who handle requested database queries. Such long waiting periods can have an adverse impact on project performance and may reduce the value of information (Ahmad and Azhar, 2005). The issue is to be able to quickly analyze existing data to discover trends so that predictions and forecasts can be made with reasonable accuracy and in a timely manner to aid in the decision-making process.

The aim of this research study is to develop a framework to guide storage and retrieval of validated and integrated data for timely decision-making and to enable construction organizations to redesign their IT infrastructure and organizational structure matched with IS capabilities. The research study examined *data warehousing*, a novel database-management technique, for its applicability and benefits that can be realized in the construction industry. Data warehousing is an improved approach for integrating data from multiple, often very large, distributed, heterogeneous databases and other information sources. It supports reorganization, integration, and analysis of data that enable users to access information quickly and accurately. The scope of this research study is limited to construction owner organizations that are continuously involved in multiple construction projects, such as transportation agencies, county/city and state governments, transit agencies, port authorities, school boards, and private corporations.

13.2 Research methodology

Construction by axiology and epistemology is a "proactive" field, in that each construction project is an intervention into what exists in the built environment and thus creates a new reality (Azhar *et al.*, 2010; Shaurette, 2013). Thus, construction research needs to employ "proactive" research approaches. Many of the researchable problems in construction tend to be applied in nature and frequently involve modifications of practices and procedures from other disciplines, and, hence, discovery of "overriding theories" is rare in construction as compared with pure engineering and natural sciences (Hauck and Chen, 1998; Ray, 1998). Hence, applied research could be more beneficial for the construction industry and may lead to better management practices, more effective field procedures, and improved levels of productivity (Azhar *et al.*, 2010). However, there are some inherent difficulties associated with applied research, such as a lack of scientific rigor, the inability to replicate the procedures, the challenge of application of results to a wider population, and the lack of dissemination owing to concerns about propriety information (Hauck and Chen, 1998). What is clearly needed in construction is a research approach that contributes towards both the solution of practical problems and creation of new theoretical knowledge. An approach that fulfils these criteria is action research. Action research produces highly reliable research results because it is grounded in practical action, aimed at solving a realistic problem situation while carefully informing theory (Baskerville, 1999).

The action research method was developed by Kurt Lewin in the wake of social changes that happened after World War II (Azhar *et al.*, 2010). He articulated the method as "a way of generating knowledge about a social system while, at the same time, attempting to change it" (Lewin, 1946). After him, scholars such as Argyris, Schön, Hult, and Lennung (see Argyris and Schön, 1978; Hult and Lennung, 1980) further developed the method.

Hult and Lennung (1980) defined action research as follows:

> Action research simultaneously assists in practical problem-solving and expands scientific knowledge, as well as enhances the competencies of the respective actors, being performed collaboratively in an immediate situation using data feedback in a cyclical process aiming at an increased understanding of change processes in social systems and undertaken within a mutually acceptable ethical framework.

Action research has three distinct characteristics that distinguish it from other research approaches (Hult and Lennung, 1980):

1   Action research aims at an increased understanding of an immediate problem situation, with emphasis on the complex and multifaceted nature of organizations.
2   Action research simultaneously assists in practical problem-solving and expands scientific knowledge. This goal extends into two important process characteristics. First, highly interpretive assumptions are made about the observations; second, the researcher intervenes in the problem setting.
3   Finally, action research is performed collaboratively and enhances the competencies of both researchers and practitioners. It links theory and practice to generate a solution.

To solve the research problem outlined in Section 13.1, the action research approach is selected for the following reasons: (1) the research portrayed a real problem situation (i.e. ineffective utilization of project data in planning and decision-making) that was identified in the construction industry through exploratory studies; (2) the research dictated the need for very close researcher–practitioner collaboration to find an adoptable solution that should satisfy the needs of the construction industry; and (3) the research domain includes construction processes, ISs, and organizational setup. Owing to the different functionalities and behaviors of these domains, the effectiveness of a solution can only be judged by its implementation and evaluation within an actual organization. Action research is one of the approaches that enables researchers to conduct such multidisciplinary studies in a real setting (Azhar *et al.*, 2010; Shaurette, 2013).

The action research process

Action research has a five-phase cyclical process. Before the cycle is commenced, it first requires the establishment of a *client-system infrastructure* or *research environment* (Susman and Evered, 1978). The client-system infrastructure is the specification and agreement that constitute the research environment. It provides the authority under which the researchers and host practitioners may specify actions. It also legitimates those actions with the express expectation that eventually these will prove beneficial to the client or host organization. Considerations found within

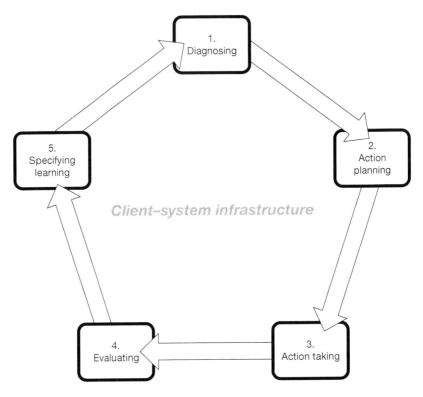

Figure 13.1 The action research cycle
Source: Adapted from Baskerville, 1999

the agreement may include the boundaries of the research domain. It may also patently recognize the latitude of the researchers to disseminate the learning that is gained in the research. A key aspect of the infrastructure is the collaborative nature of the undertaking. The researchers work closely with practitioners who are located within the client system. These individuals provide the subject system knowledge and insight necessary to understand the anomalies being studied (Clark, 1972). Once the client-system infrastructure is established, the five-phased cycle of action research commences. As shown in Figure 13.1, the phases are:

Phase 1: Diagnosing or problem identification

Diagnosing corresponds to the identification of the primary problems that are the underlying causes of the organization's desire for change. Diagnosing involves self-interpretation of the complex organizational problem, not through reduction and simplification, but rather in a holistic fashion. This diagnosis leads to the development of certain theoretical assumptions (i.e., a working hypothesis) about the nature of the organization and its problem domain (Baskerville, 1999).

Phase 2: Action planning

Researchers and practitioners then collaborate in the next activity, action planning. This activity specifies organizational actions that should relieve or improve these primary problems. The discovery of the planned actions is guided by the theoretical framework, which indicates both the desired future state for the organization and the changes that would achieve such a state. The plan establishes the target for change and the approach to change (Baskerville, 1999).

Phase 3: Action taking

The planned actions from Phase 2 are implemented in the action-taking phase. The researchers and practitioners collaborate in the active intervention into the client organization, causing certain changes to be made. Several forms of intervention strategy can be adopted. For example, the intervention might be directive, in which the research "directs" the change, or non-directive, in which the change is sought indirectly. Intervention tactics can also be adopted, such as recruiting intelligent laypersons as change catalysts and pacemakers. The process can draw its steps from social psychology, e.g., engagement, unfreezing, learning, and reframing (Baskerville, 1999).

Phase 4: Evaluating

After the actions are completed, the collaborative researchers and practitioners evaluate the outcomes. Evaluation includes determining whether the theoretical effects of the action were realized, and whether these effects relieved the problems. Where the change was successful, the evaluation must critically question whether the action undertaken, among the myriad routine and non-routine organizational actions, was the sole cause of success. Where the change was unsuccessful, some framework for the next iteration of the action research cycle (including adjusting the hypotheses) should be established (Baskerville, 1999).

Phase 5: Specifying learning

Although the activity of specifying learning is formally undertaken last, it is usually an ongoing process. The knowledge gained in the action research (whether the action was successful or unsuccessful) can be directed to three audiences (Baskerville, 1999): (1) the restructuring of organizational norms to reflect the new knowledge gained by the organization during the research; (2) where the change was unsuccessful, the additional knowledge may provide foundations for diagnosing in preparation for further action research interventions; and (3) the success or failure of the theoretical framework provides important knowledge to the scientific community for dealing with future research settings.

The action research cycle can continue, whether the action proved successful or not, to develop further knowledge about the organization and the validity of relevant theoretical frameworks. As a result of the studies, the organization thus learns more about its nature and environment, and the constellation of theoretical elements of the scientific community continues to benefit and evolve (Argyris and Schön, 1978).

Demonstration of the action research process

Based on the action research approach, a five-step methodology is designed for the research problem defined above. Figure 13.2 illustrates this methodology.

A brief outline of the research methodology is as follows:

1   assessment of data-management practices and level of utilization of ISs for planning and decision-making in construction owner organizations;
2   organizational, functional, information, and decision modeling of selected owner organization;
3   development of functional model and reference architecture using data-warehousing technique; it involves development of data extraction, validation and integration schemas, multidimensional data models, and OLAP query techniques;

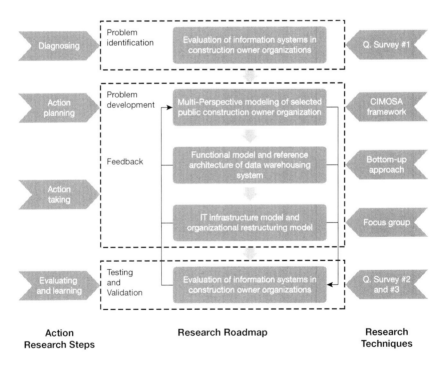

Figure 13.2 The research methodology framework
Source: Azhar, 2005

4   development of the decision-support framework showing different organizational setups and matching IT infrastructure to effectively utilize project data for decision-making; and

5   validation of the decision-support framework; both qualitative and quantitative measures are used for validation: qualitative measures include improvements in data access, data quality, productivity, response time, etc., and quantitative measures consist of time-savings and operational cost savings.

13.3 Data-collection strategy

Before any data are collected, the researcher should have the following information: (1) Who will be the collaborator(s)? (2) What are the responsibilities and tasks of the researcher and the client organization? (3) And who has the right to knowledge generated (prior, during, after) and the intellectual property rights? Typically, the researcher and the collaborator (or client organization) would enter into a formal contract before any data are collected. Depending on the type of research problem, the data could be collected from a single organization (if the research problem is diagnosed in a single organization) or from multiple organizations (if several organizations are facing the same problem). In the latter case, typically the researcher chooses one of the identified organizations as a "collaborator/client organization," and the remaining organizations are used for solution-verification and -validation purposes. Depending on the nature of the problem, the required data could be collected using quantitative, or qualitative, or mixed methods approaches.

In line with the research methodology framework depicted in Figure 13.2, the first step in our research study was to evaluate the existing data-management practices in construction owner organizations and the level of utilization of ISs in planning and decision-making (i.e. the problem identification phase). The construction owner organizations were chosen in this research study for three reasons: (1) owners have a wider perspective as they are involved with the project from inception to completion and, in most cases, are the ultimate users of the constructed facilities; (2) owners are in control of project funds; the extent of this control, however, depends on the type of contractual arrangement, but owners are the ultimate stakeholders, as far as the overall investment in a project is concerned; and (3) owner organizations' (public agencies', in particular) data are easily available to the public. In the next step (i.e., the action planning and action taking phases), one local owner organization (hereinafter called the client organization) was chosen as a collaborator. The client organization is in charge of administration of all public transportation-related construction projects in the Miami-Dade County of the State of Florida. The reasons for this selection were its staff's motivation about this research, firm commitment of executive management, and grant of access to research-related data. In the last step (i.e., the evaluation and learning phase), the "solution" (i.e., the final prototype system) was verified and validated to ensure that the research objectives had been achieved. The verification and validation were conducted both *internally* and *externally*. The internal validation was performed by members of the focus group. For external validation, twenty construction owner organizations that were shortlisted from Phase 1 (i.e., the problem identification phase) were used.

13.4 Data–collection methods

The following data-collection methods are typically common in action research:

1   *observations* (i.e., own reflections such as diary writing, it is an intuitive process that allows individuals to collect information about others by viewing their actions and behaviors in their natural surroundings);
2   *individual interviews* (others' interpretations and reflections to get the individual stories);
3   *focus groups* (group discussions among experts to reach a consensus); and
4   *questionnaires* (to measure attitudes, factors, or values held by individuals).

Readers interested to find out more details about these data-collection methods are advised to read McClure (2002) or any other text on research methods.

The data-collection methods used in the different phases of the presented case study are summarized below:

Phase 1: Problem identification

The research need was established on a hypothesis that construction owner organizations do not effectively utilize project data for planning and decision-making owing to a lack of decision support in their existing ISs. To validate the hypothesis, a questionnaire survey was conducted. The questionnaire was sent to 550 construction owner organizations, and responses were collected in a period of 3 months. The questionnaire comprised fifteen questions that were grouped into three sections, namely (1) organization profile; (2) evaluation of data-management practices and degree of utilization of ISs in planning and decision-making operations; and (3) need for data warehousing in the surveyed organizations. A fourth optional section was provided at the end to collect personal information about the respondents. The questionnaire was prepared in two formats: an online or web-based format and a printed or hard-copy format. Reminders were sent 30 days and 60 days after the questionnaire was sent out. Of the 550 questionnaires sent, 163 valid responses were received, representing a total response rate of approximately 30 percent.

Phases 2 and 3: Action planning and action taking

Three main tasks, namely: (1) enterprise modeling, (2) prototype system develop-ment, and (3) restructuring of IT infrastructure and organizational setup, were performed in these phases.

The purpose of enterprise modeling was to capture the current or "as-is" state of the client organization and model it from various perspectives, such as functional, organizational, informational, decision-making, etc. These models were developed to understand the organization's business operations, the flow of information within, into, and outside the organization, identification of decision nodes and their hierarchy, and data requirements for these decisions. Relevant data were collected using observations, individual interviews, and review of project documents.

Based on the user requirements identified in the previous task, a functional (or logical) model of the prototype system and its corresponding reference architecture were developed. The functional model represents how the system would work, and the reference architecture indicates its physical and technical implementation using different software and hardware tools. A focus group consisting of the following six executives of the client organization was formed to assist in the data-collection process and to test the prototype system: manager, contracts administration; manager, cost and scheduling; senior manager, construction management services; manager, design and engineering; manager, computer services; and manager, information technology and support services.

In consultation with the focus-group members, the researchers decided to test the data-warehousing technology as a promising solution to solve the problems diagnosed in the first phase. A data warehouse is a dedicated database system created by combining data from multiple databases for purposes of analysis. It organizes the collected data for consistency and easy interpretation, keeps "old" data for historical analysis, and makes access to, and use of, data a simple task, so that users can do it themselves without great technical proficiency in data handling (Ahmad and Azhar, 2002). Its purpose is to provide easy access to especially arranged data that can be used with decision-support applications, such as management reporting, queries, data mining and executive ISs (Ahmad and Azhar 2005).

During the course of the research, it was recognized that the successful implementation of the prototype system would require appropriate changes in the IT infrastructure and organizational structure of the client organization to maximize productivity and ensure maximum cost savings. The functional model and the prototype system were shown to the focus group to explain their functionalities and possible role in the future decision-support operations. After the demonstration, the group gave its feedback on how the implementation of the full-scale system would affect the existing IT infrastructure and organizational structure. Based on this input, several IT infrastructure and organizational redesign schemas were developed. After a series of discussions, the focus group approved a final IT infrastructure and organizational restructuring model for implementation. Finally, the prototype system was implemented in the client organization to assess its effectiveness. Training workshops were conducted to educate the members of the focus group about this new system.

Phases 4 and 5: Evaluating and learning

In the final two phases, the focus-group members tested the performance of the system, verified its accuracy, and validated its effectiveness using several quantitative and qualitative measures. The validation process was conducted both *internally* and *externally*. The questionnaire survey approach was adopted for validating the system in both internal and external phases. The questionnaire consisted of fourteen ranking questions divided into three sections, as follows: (1) improvements in data-management operations; (2) improvements in decision-support operations; and (3) cost and time-savings in various planning and decision-making operations. Before sending out the questionnaire survey, the respondents in the twenty shortlisted organizations were contacted via phone. They briefly had explained the nature of the research, potential findings, and their role in the validation process. After their agreement, the questionnaire survey and a video illustrating the different stages of the prototype system and the resulting framework were sent. A week after dispatch of the surveys, all respondents were again contacted via telephone so that any queries could be answered. It was ensured that all respondents had correctly understood the concept and functionalities of the prototype system. This step was necessary to eliminate any bias in results that might develop owing to different interpretations of the prototype system by the respondents. Based on the feedback of focus-group members and all respondents, minor changes were made in the prototype system. Readers interested in more details should consult Azhar (2005, 2008).

13.5 Data analysis and results

The collected data in action research can be analyzed using quantitative (e.g. statistical) or qualitative methods (using tools such as HyperResearch®, Atlas®, Vivo®, etc.). Typically, the research results are presented in two formats: an expert report for the client organization and a research report for the research organization, such as dissertation, thesis, or a scholarly article.

The research results of our case study are broken down into four sections as follows.

Phase 1: Validating the research needs

This research study was based on a hypothesis that the construction owner organizations do not effectively utilize project data for planning and decision-making. The results of the first questionnaire survey validated this hypothesis. The results indicated that only 40 percent of surveyed organizations employ a formal database management system to store and manage the project data. It was also found that 39 percent of these organizations use their existing ISs to make everyday decisions, whereas only 29 percent use them for short- and long-term planning. The reasons found were low decision-support capabilities of existing ISs, poor data quality, difficult data access, and high cost of operations. Ninety-one percent of surveyed organizations indicated that they strongly need a new and enhanced IS to support planning and decision-making operations.

Phases 2 and 3: Planning and implementation of the prototype system

After the prototype system was designed and implemented in the client organization, the focus-group members tested the performance of the system and assessed its effectiveness. During this stage, members of the focus group deeply analyzed the organization's existing IT infrastructure and organizational structure and developed an overhaul plan to maximize the benefits. The following subsections highlight the main findings.

The new IT infrastructure model to support the prototype system

The existing IT infrastructure of the client organization, like many other construction owner organizations, has not been planned and developed, along with eventual "computerization" of different units and processes, during the last two to three decades. This IT infrastructure can be termed as "application-centric," meaning that separate applications are developed to run various discrete processes, and, as a result, data needed by individual processes are stored in individual applications. There is minimal integration among these applications, and users often have to manually integrate data from different applications to generate information needed for the purpose of decision-making. To implement the data-warehousing-based prototype system, the current IT infrastructure needs to be redesigned to allow real-time, two-way communications between the users (or data sources) and the data warehouse. The new IT infrastructure, which is based on the center-to-center (C2C) communications concept, is shown in Figure 13.3. C2C is a two-way communication approach that means that all units in an organization are interconnected via a network and periodically supply data to the data warehouse, which stores it in a central repository (Watson, 2001).

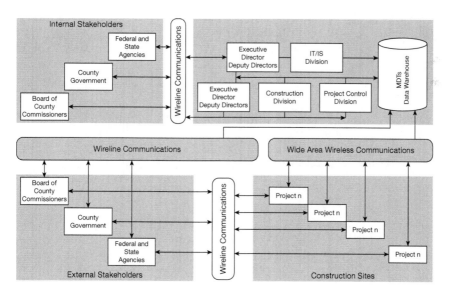

Figure 13.3 New IT infrastructure for the client organization

Source: Concept from Courage *et al.*, 2004

The organizational restructuring model to maximize the effectiveness of the prototype system

The information management capabilities of an organization strongly influence its design (Nosek, 1989). Organizational design is considered a dynamic process. A number of research studies indicated that organizational design and the design of work processes is shaped by the amount and type of information required in a given environment and the organization's information processing capability (Galbraith, 1974; Tushman and Nadler, 1978). In construction organizations, the organizational structure is often rigid and is usually not redesigned after the implemention of new or enhanced ISs. This situation often results in higher operational costs and lower productivity because the organizational structure and ISs are not compatible with each other. Data warehouses have the potential to support significant changes in how an organization is structured and carries out its business. Because a data warehouse can provide more detailed, integrated, and historically complete information, it should be possible for an organization to operate very differently and "restructure" itself (Watson *et al.*, 2001).

After testing the prototype system, the focus group proposed that the organizational restructuring should be carried out after 1 year of full-scale system implementation and in two phases. In the first phase, the construction division and project-control division of the client organization would be downsized, and,

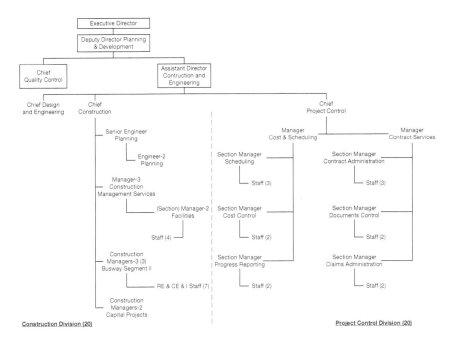

Figure 13.4 Existing organizational structure of the client organization

Source: Azhar, 2005

in the second phase, they would be merged to form a new division. These organizational changes would help to decrease the hierarchy in the organizational structure and result in substantial cost savings. The existing organizational structure of the construction and project-control divisions is shown in Figure 13.4.

Initially, there were forty employees working in these two divisions. During the group discussions, it was determined that most of the computational work was being performed by the section managers. They collect project data from the office and field staff, organize and analyze the data, prepare reports, and send these reports to respective mangers or division chiefs for necessary decision-making. After implementation, these roles would be carried out by the data-warehousing system (by making appropriate changes in the business processes), and the services of section managers in the existing form may no longer be required. Hence, it was decided to eliminate these positions after 1 year of full-scale system implementation. The new organizational structure is illustrated in Figure 13.5. The significant features of this organizational structure are reduced organizational hierarchy and downsizing of management staff by 23 percent. It was also decided that the data-warehousing system would be maintained by the existing staff of the IT division (not shown in Figures 13.4 and 13.5). It was further decided that this new organizational structure should persist for a period of 2 years (3 years in total, after including 1 year of

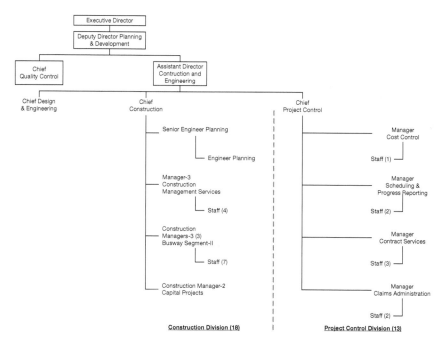

Figure 13.5 Proposed organizational structure after 1 year of data warehousing implementation

Source: Azhar, 2005

training). During this time, it was expected that the division chiefs and managers would have developed necessary expertise to use the data-warehousing system for their daily, short-term and long-term business needs. At the same time, the data warehouse would have been populated with complete historical and current project data. Only after these milestones had been met should the second phase of organizational restructuring be carried out.

During Phase 1, it was assessed that different functional areas of project management had been split between the two divisions, and, as a result, unnecessary time and effort were wasted in coordination and decision-making. The various processes could be streamlined if both divisions were merged to work under a single management. Based on this premise, the second phase of organizational restructuring was planned, as shown in Figure 13.6. It was proposed to merge the construction division and project control division to form a new "construction planning and management division." This new division would carry out all operations related to construction planning, management, and control. As the division would work under a single division chief, the decision-making would be quicker and more effective. It was further expected that this change would increase coordination between the employees and result in improved productivity. The number of

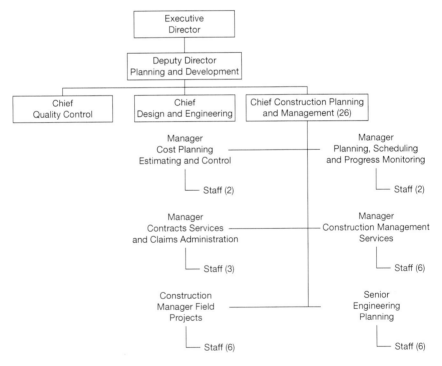

Figure 13.6 Proposed organizational structure after 3 years of data warehousing implementation

Source: Azhar, 2005

employees would now be reduced to twenty-six, thereby indicating a 35 percent reduction in workforce compared with the existing one.

Phase 4: Evaluating the prototype system

For evaluation (i.e., verification and validation) purposes, limited data were fed into the prototype system. The results were compared with the existing project reports. It was found that the variances between the prototype system results and those prepared by using manual calculations or customized spreadsheets were less than ±1 percent. The comparison of results verified that the logics and schemas used in the development of the prototype system were correct. Improvements in the decision-making processes were measured by the focus-group members (i.e., the internal validation) using various performance measures. Table 13.1 shows the focus-group opinions before and after using the prototype system.

The results indicate that the significant error (p) in all cases is below 0.05 (< 5 percent). This means that the results are statistically significant: that is, significant improvements were recorded after the implementation of the prototype system. Further data analysis indicated that aapproximately 20–30 percent operational time-savings and 10–20 percent operational cost savings could be realized through data warehousing. Approximately similar results were obtained through external validation. Interested readers can find out these results in Azhar (2005).

Table 13.2 compares the costs incurred for data-warehousing implementation with the resulting direct cost savings that would be achieved as a result of changes in the organizational structure. The net present value method is used to convert all future cost savings into present-day dollars (assuming 6 percent discount rate). As is evident from Table 13.2, a period of approximately 3.5–4 years would be required to pay off the data-warehousing costs by means of savings in direct

Table 13.1 Comparison of performance before and after implementation of the prototype data–warehouse system

Comparison criteria	Mean of responses		Sum of squares (X^2)	Significant error (p)
	Existing methods	*Prototype system*		
Ease of data access	2.83	4.33	8.42	0.01
Data quality	3.12	4.67	6.97	0.03
Data integration	2.45	4.56	7.01	0.02
Organizational productivity improvement	1.98	4.67	9.14	0.01
Quality of reports	2.78	3.67	5.12	0.03
Support for every day decisions	2.46	4.67	10.12	0.00
Support for short- and long-term planning	1.78	4.83	6.98	0.01

Source: Azhar, 2005

Table 13.2 Comparison of data-warehousing implementation cost and net associated direct cost savings

Time after data-warehousing implementation	Initial and operational costs ($)	Savings due to organizational restructuring ($)	Net savings ($)
3 years	1,231,639	795,917	435,722
5 years	1,691,832	3,092,461	1,400,629
10 years	2,465,429	11,090,792	8,625,363

Source: Azhar, 2005

salary costs. After 5 and 10 years, net savings in direct salary costs would be approximately $1.4 million and $8.6 million, respectively. It is important to note that data warehousing in itself would result in substantial savings in indirect costs, such as information-management and -processing costs, report-preparation costs, and other miscellaneous operational costs. However, these costs are not included here because the focus of this comparison is to show the benefits of organizational restructuring on direct cost savings.

Phase 5: Specifying learning (the final decision-support framework)

The end product of this research is a five-phase framework for planning, designing, building, and implementing a data-warehousing systems in construction owner organizations (see Figure 13.7). As the nature and sequence of construction operations in most construction owner organizations are same, and the decision-making requirements are not much different, this framework and its associated models can be used by other organizations with few modifications. These organizations do not have to "reinvent the wheel." This effort will save a significant amount of time and resources for these organizations.

The final deliverables of this research study included the following: (1) a detailed report for the client organization highlighting the solution (i.e., the prototype system) and its implementation plan; (2) a Ph.D. dissertation; (3) several scholarly publications in peer-reviewed conferences and journals (see Ahmad and Azhar, 2005; Azhar and Ahmad, 2007; Azhar, 2008; Azhar *et al.*, 2010). In a nutshell, this research study helped to build and test theory while solving a realistic problem within an actual organization.

13.6 Main findings and conclusions

It is demonstrated through the case study that action research provides a structured approach to conduct applied research while maintaining a high level of academic rigor and permitting the application of results to a wider audience. It is highly suitable for conducting applied research in construction, especially in multidisciplinary research that simultaneously involves technological, organizational, and behavioral

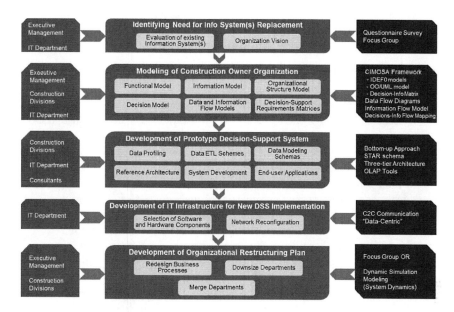

Figure 13.7 A framework for data-warehouse design and implementation in construction owner organizations

Source: Azhar, 2005

aspects. Examples of such areas are the design and testing of new equipment, systems, or business processes, and the design and implementation of ISs (Azhar *et al.*, 2010). Action research can also effectively help to improve collaboration between academic researchers and industry practitioners in research and development projects (Azhar, 2007). Action research addresses a specific problem situation, although it generates knowledge that enhances the development of general theory. The domain of action research method is characterized by a social setting where: (1) the researcher is actively involved, with expected benefits for both researcher and organization; (2) the knowledge obtained can be immediately applied; there is not the sense of a detached observer, but that of an active participant wishing to utilize any new knowledge based on an explicit, clear, conceptual framework; and (3) the researcher wants to link theory and practice to generate a solution. One clear area of importance in the domain of action research is new or changed systems-development methodologies. Studying new or changed methodologies implicitly involves the introduction of such changes and is necessarily interventionist. From a social–organizational viewpoint, the study of a newly invented technique is impossible without some type of intervention to inject the new technique into the practitioner environment: that is, "go into the world and try them out" (Wood-Harper, 1989). Action research is one of the few valid research approaches that can legitimately be employed to study the effects of specific alterations in systems-development methodologies in human organizations (Baskerville and Wood-Harper, 1996).

Action research processes and typical organizational consulting processes contain substantial similarities, and it is easy to confuse the two approaches. However, action research and consulting differ in five key ways (Lippitt and Lippit, 1978; Kubr, 1986):

1   *Motivation*: Action research is motivated by its scientific prospects, perhaps epitomized in scientific publications. Consulting is motivated by commercial benefits, including profits and additional stocks of proprietary knowledge about solutions to organizational problems.
2   *Commitment*: Action research makes a commitment to the research community for the production of scientific knowledge, as well as to the client. In a consulting situation, the commitment is to the client alone.
3   *Approach*: Collaboration is essential in action research because of its idiographic assumptions. Consulting typically values its "outsider's" unbiased viewpoint, providing an objective perspective on the organizational problems.
4   *Foundation for recommendations*: In action research, this foundation is a theoretical framework. Consultants are expected to suggest solutions that, in their experience, proved successful in similar situations.
5   *Essence of the organizational understanding*: In action research, organizational understanding is founded on practical success from iterative experimental changes in the organization. Typical consultation teams develop an understanding through their independent critical analysis of the problem situation.

In summary, consultants are usually paid to dictate experienced, reliable solutions based on their independent review. Action researchers act out of scientific interest to help the organization itself learn by formulating a series of experimental solutions based on an evolving, untested theory (Baskerville, 1997).

Despite all its advantages, the action research approach has some deficiencies. For example, the collaborative research framework could potentially diminish the researcher's ability to control the process and the outcomes of the research. Another potential problem can arise from the need for researchers and practitioners to share a mutually acceptable ethical framework. Successful action research is unlikely where there is conflict between researchers and practitioners, or among practitioners themselves. For example, problems may well arise if the research could lead to people being fired (Avison *et al.*, 1999). Such a result can be in conflict with the researcher's principles, but be acceptable to practitioners, or vice versa (Azhar *et al.*, 2010). Nevertheless, action research provides a rewarding experience for researchers who want to work closely with industry. It can be combined with other research methods to generate new theory and/or to reinforce or contradict existing theory. Similarly, it can benefit the industry by evolving improved management practices, more effective field procedures, and development of new products, materials, and tools, which could result in high levels of tangible and intangible benefits (Azhar *et al.*, 2010). In that way, action research can effectively help to diminish the perception among many construction practitioners that academic research is less relevant to the industry and more focused on subjects and issues that are not crucial for the construction industry.

References

Ahmad, I. and Azhar, S. (2002). "Data Warehousing in Construction: From Conception to Application." *Proceedings of the First International Conference on Construction in the 21st Century: Challenges and Opportunities in Management and Technology*, April 24–26, Miami, Florida, USA, pp. 739–747.

Ahmad, I. and Azhar, S. (2005). "Implementing Data Warehousing in the Construction Industry: Opportunities and Challenges." *Proceedings of the Third International Conference on Construction in the 21st Century (CITC-III)*, September 15–17, Athens, Greece, pp. 863–871.

Ang, J. and Teo, T. S. H. (2000). "Management Issues in Data Warehousing: Insight from the Housing and Development Board," *Decision Support Systems*, 29: 11–20

Argyris, C. and Schön, D. (1978). *Organizational Learning: A Theory of Action Perspective*. Reading, MA: Addison-Wesley.

Avison, D., Lau, F., Myers, M., and Nielsen, P. A. (1999). "Action Research," *Communication of the ACM*, 42(1): 94–97.

Azhar, S. (2005). *Information Systems to Support Decision-Making in Construction Owner Organizations: A Data Warehousing Approach*. Ph.D. dissertation, Florida International University, Miami, Florida.

Azhar, S. (2007). "Improving Collaboration between Researchers and Practitioners in Construction Research Projects using Action Research Technique." *Proceedings of the 43rd ASC National Annual Conference*, April 12–14, Flagstaff, AZ [on CD-ROM].

Azhar, S. (2008). *Data Warehousing in Construction Organizations: Concepts, Architecture and Implementation*. Saarbrücken: Germany: VDM Verlag Muller.

Azhar, S. and Ahmad, I. (2007). "A Case Study of Data Warehousing Implementation in a Public Construction Owner Organization." In S.-H. Song (ed.), *Construction Data Warehouse* (2nd edn, pp. 15–31). Seoul, Korea: Seoul National University.

Azhar, S., Ahmad, I., and Sein, M. K. (2010). "Action Research as a Proactive Research Method for Construction Engineering and Management," *ASCE Journal of Construction Engineering & Management (Special Issue on Research Methodologies in Construction Engineering & Management*, eds J. E. Taylor and E. J. Jaselskis), 136(1): 87–98.

Baskerville, R. (1997). "Distinguishing Action Research from Participative Case Studies." *Journal of Systems and Information Technology*, 1(1): 25–45.

Baskerville, R. (1999). "Investigating Information Systems with Action Research," *Communications of the Association of Information Systems*, 2(19): 7–17.

Baskerville, R. L. and Wood-Harper, A. T. (1996). "A Critical Perspective on Action Research as a Method for Information Systems Research," *Journal of Information Technology*, 11: 235–246.

Clark, P. (1972). *Action Research and Organizational Change*. London: Harper & Row.

Courage, K., Li, I., Hammer, J., Ji, F., and Yu, Q. (2004). *Feasibility Study for an Integrated Network of Data Sources*, Final report. University of Florida, Gainesville and Florida Department of Transportation, January.

Galbraith, J. R. (1974). "Organizational Design: An Information Processing View," *Interfaces*, 4(3): 87–89.

Hauck, A. and Chen, G. (1998). "Using Action Research as a Viable Alternative for Graduate Theses and Dissertations in Construction Management," *International Journal of Construction Education & Research*, 3(2): 79–91.

Hult, M. and Lennung, S. A. (1980). "Towards a Definition of Action Research: A Note and Bibliography," *Journal of Management Studies*, 5(2): 241–250.

Kubr, M. (1986). *Management Consulting: A Guide to the Profession* (2nd edn). Geneva: International Labour Office.

Lewin, K. (1946). "Action Research and Minority Problems," *Journal of Social Issues*, 2(4): 34–46.

Lippitt, G. and Lippit, R. (1978). *The Consulting Process In Action*. San Diego, CA: University Associates.

McClure, R. D. (2002). "Common Data Collection Strategies Effective in Qualitative Studies using Action Research in Technical/Operational Training Programs [online]." Available at http://evokedevelopment.com/uploads/blog/commonData.pdf (accessed December 4, 2015).

Naoum, S. G. (2001). *Dissertation Research and Writing for Construction Students*. Oxford, UK: Butterworth Heinemann.

Nosek, J. T. (1989). "Organization Design Strategies to Enhance Information Resource Management," *Information & Management*, 16: 81–91.

Ray, C. S. (1998). "An Action Research Plan for Developing and Implementing Writing Skills in Construction Project Administration." *Proceedings of the 34th Annual International ASC Conference*, April 15–8, New Britain, CT.

Rezgui, Y. (2007). "Exploring Virtual Team-Working Effectiveness in the Construction Sector," *Interacting with Computers*, 19(1): 96–112.

Shaurette, M. (2013). "Using Action Research to Refine Demolition Curriculum for Construction Management Students." *Proceedings of the 45th Annual International ASC Conference*, April 1–4, Gainesville, FL.

Susman, G. and Evered, R. (1978). "An Assessment of the Scientific Merits of Action Research," *Administrative Science Quarterly*, 23(4): 582–603.

Tushman, M. L. and Nadler, D. A. (1978). "Information Processing as an Integrating Concept in Organizational Design," *Academy of Management Review*, 3: 613–24.

Watson, H. J. (2001). "Designing Data Warehouses," *Data & Knowledge Engineering*, 31: 279–301.

Watson, H. J., Goodhue, D. L., and Wixom, B. H. (2001). "The Benefits of Data Warehousing: Why Some Organizations Realize Exceptional Payoffs," *Information & Management*, 1–12.

Wood-Harper, T. (1989). *Comparison of Information Systems Definition Methodologies: An Action Research Multiview Perspective*. Ph.D. thesis, University of East Anglia, UK.

Part VI

Grounded theory research

14 The application of grounded theory methodology in built environment research

Monty Sutrisna and Wisnu Setiawan

The built environment has established itself as a distinctive academic discipline with its unique research traditions and practice. Following the call from its scholars to further enrich the choice and utilisation of more diverse research methods, tools, and techniques, built environment research has proven to be fertile ground for the implementation and adaptation of a vast array of research approaches. This includes bringing research techniques that have been previously considered 'less traditional' for the built environment discipline, including the grounded theory methodology. Stemming from the symbolic interaction tradition, the grounded theory methodology has been successfully implemented in various built environment research. The very capability of its data-collection technique in yielding rich data and the robustness of its data-analysis techniques in deriving meaningful research findings have made the grounded theory methodology a popular choice in the contemporary built environment research arena. This chapter discusses the background of grounded theory in research methodology, grounded theory in research design, data collection and data analysis in grounded theory, as well as different ways of implementing grounded theory in research projects. Implementation can range from the full-scale implementation of the grounded theory methodology as a holistic research methodology to the adaptation of the grounded theory methodology as a stand-alone data-analysis technique. The implication of different levels of adaptation towards the overall research methodology in, and impact towards, the final findings of these different research projects is discussed here. A reflection section is included to highlight lessons learned, as well as to discuss the way forward in unlocking further potential of the grounded theory methodology in built environment research.

14.1 Introduction

By incorporating a range of specialisations relevant to the design, development and management of buildings, spaces and places (Griffiths, 2004), the built environment has emerged as a distinctive academic discipline with its unique research practices and traditions. While developing further to expand its theories and applications, research studies in the built environment have been criticised for the academic drift from its practice-based origins (Chynoweth, 2013), so much so that, in many cases, practitioners failed to appreciate the value and benefits of academic research (Gann, 2001). Similar to other disciplines achieving this level of maturity, built environment research has 'reached a stage that demands the validation of its principles within different real world situations in order to refine and integrate them' (Amaratunga *et al.*, 2002: 17). In an attempt to refine and integrate research in the built environment discipline, scholars have been debating about the most appropriate research approaches (e.g. Raftery *et al.*, 1997; Rooke *et al.*, 1997; Runeson, 1997; Seymour *et al.*, 1997) and calling for further diversity in research method application (e.g. Dainty, 2008). In response to this, researchers have used built environment research as a ground for the implementation and adaptation of a vast array of research approaches. This includes the adaptation of research techniques from other disciplines that may have been previously considered 'less traditional' for the built environment discipline, including the grounded theory methodology.

The grounded theory methodology has been considered a suitable technique to be applied when one is investigating a substantive area about which little is known, or about which much is already known, but novel understanding is desired (Stern, 1980). Generally speaking, the grounded theory methodology aims to develop theory that is grounded in systematically gathered and analysed data (Crotty, 1998; Strauss and Corbin, 1998; Groat and Wang, 2002). Scholars have reported difficulties in adopting the grounded theory methodology, including data overload, over-complex procedures, the lengthy analytical phase. the existence of researchers' assumptions, personal beliefs and values or 'ideological baggage', and so on (e.g. Charmaz, 1990; Dainty *et al.*, 2000; Hunter and Kelly, 2008). Despite these difficulties, the grounded theory methodology has been reported to have been successfully implemented in built environment research (e.g. Loosemore, 1999; Dainty *et al.*, 2000; Barrett and Sutrisna, 2009) and considered appropriate on many occasions to assist understanding of humans' experience in a rigorous and detailed manner (Ryan and Bernard, 2003). The development of 'substantive theory' using the grounded theory methodology is typically facilitated by distillation of data through open, axial, selective coding guiding the researcher by indicating avenues of further investigation to conduct further 'cycles' of data collection and analysis (Strauss and Corbin, 1998). The typical use of the (multiple) case study approach has been recognised to be in line with the principles of the grounded theory methodology at the methodological level, that is, using multiple sources and applying constant comparison of different case studies, representing the 'cycles', to achieve the point of saturation where further data collection and the constant comparison within and among the case studies cease to yield new information (Sutrisna and Barrett, 2007).

This chapter aims to also present some examples of implementation of the grounded theory methodology in built environment-related research projects within various settings and context, by analysing its implementation reported in those research. There is evidence suggesting the potential of implementation, ranging from full-scale implementation of the grounded theory methodology as a holistic research methodology to the adaptation of the grounded theory methodology as a stand-alone data-analysis technique, as well as stand-alone data-collection techniques and procedures. The methodological soundness of these 'partial' implementations of the grounded theory methodology and the implication of these different levels of adaptation for the overall research methodology in, and its impact towards, the final findings of these different research projects are discussed in this chapter. After considering the advantages and the potential downside of adopting, it can be concluded that the different levels of grounded theory methodology implementation can most likely be beneficial for the research. Although the benefit appears to outweigh the difficulties associated with its implementation, there are a few things that researchers must be aware of before they decide to adopt grounded theory in their research project. It is the intention of the authors of this chapter to provide guidance for the implementation of the grounded theory methodology in built environment research.

14.2 Grounded theory in research methodology

Prior to the design of the research methodology, there is a need for the researcher to clarify the underlying meta-theoretical assumption (Crotty, 1998; Del Casino *et al.*, 2000; Barrett and Barrett, 2003; Sutrisna, 2009). In doing so, researchers are typically clarifying their ontological and epistemological stances. Whereas ontology is concerned with the nature of reality, epistemology is concerned with human access to reality. As one of the epistemological stances, constructionism views meaning (or knowledge) as continuously constructed through the interaction between the human mind and the reality, rather than 'discovered' by the human mind, as suggested in the positivists' view (Crotty, 1998). The epistemological stance of constructionism forms the basis for interpretivist theoretical perspectives, with symbolic interactionism being one of these perspectives (Crotty, 1998; Schwandt, 1998). Symbolic interactionism views practical application as the main principle to determine truth and values, to the extent of accepting uncritical exploration of ideas and values and their practical outcomes to make the reality directly accessible to the observer (Denzin, 1993; Mingers, 2004). One of the research techniques that stem from the symbolic interactionism tradition is the grounded theory methodology, which was originally introduced by Glaser and Strauss (1967). Although it is gaining popularity as a credible, plausible research technique, it is noted that many researchers using grounded theory have failed to adequately discuss both their epistemological and ontological stance and quality and validity in their research (Lomborg and Kirkevold, 2003). Therefore, the first contribution made by this chapter is to encourage built environment researchers to clearly discuss their ontology and epistemology stance prior to making a decision to adopt a particular technique, such as the grounded theory methodology.

Originally developed by Glaser and Strauss (1967), grounded theory can be considered a general methodology intended to develop substantive theory that is grounded in collected and analysed data (Glaser and Strauss, 1967; Strauss and Corbin, 1998; Cutcliffe, 2000; Sutrisna and Barrett, 2007). Grounded theory aims to develop theory through continuous interplay between analysis and data collection. Hence, theories emerging from the grounded theory methodology always correspond to their original source of data, representing the interaction among the actors incorporating its temporality and process. This continuous and consistent comparison of theory with the captured data results in a strong relationship between theory and data, hence the typically strong practical usefulness of the research findings (Strauss and Corbin, 1998). Further development of the grounded theory methodology witnessed a divergence of the principle and technique into Glaserian (Glaser, 1978, 1999), Straussian (Strauss and Corbin, 1998) and Charmaz's constructivist grounded theory (Charmaz, 2000).

Glaserian grounded theory is characterised by full dedication to the actual data collected and the expectation that researchers will be free of any preconceived ideas about the matters being researched. Straussian grounded theory, on the other hand, recognises the role of the researcher in interpreting the data, who may well be influenced by prior knowledge regarding the subject being researched. Thus, keeping up with the symbolic interactionism tradition, Straussian grounded theory put more emphasis on practical applications of conducting research, resulting in a reputation as the 'more flexible' approach compared with Glaserian grounded theory. Charmaz's constructivist grounded theory perceived the research as a dynamic discourse between the researcher's and the respondent's view of the world (reality). Intuitively, researchers may come up with the question of whether it will be possible to combine some aspects of these different 'directions' of grounded theory, with a view to selecting the 'best bits' to tackle their research problems. Even though all of these different 'directions' follow inductive reasoning as the major common denominator, the differences between them are mainly philosophical ones (refer to Lomborg and Kirkevold, 2003, for further discussion on this). Therefore, it should be recommended that researchers do not try to 'mix' them. This can be strongly related to the encouragement of built environment researchers to stay "truthful" to their belief about the nature of reality and how researchers should make sense of the assumed reality (Sutrisna, 2009).

14.3 Research design in grounded theory

One of the most important outcomes of research design is the determination of research components/activities and the sequencing of these research activities. When research is designed, the typical considerations involve purpose, conceptual framework, research question, methods and sampling procedures. To achieve purpose, facilitate the development of a conceptual framework and provide answers to the research questions, the typical research components/activities include literature review, data collection and data analysis. Data analysis will only occur subsequent to data collection. Therefore, the main consideration here would be the sequencing between literature review and data collection. The sequencing of these two major components will be heavily influenced by the research project's reasoning, which (without discounting other reasoning approaches) can generally be grouped as deductive or inductive reasoning.

In deductive reasoning, data collection has the purpose of testing a hypothesis and will typically be conducted subsequent to the development of the hypothesis, based on an in-depth literature review. Data collection will typically occur earlier in an inductive research, using techniques such as grounded theory, owing to the important role that data collection plays in developing the research outcome. Data collection in grounded theory holds a prominent position in the development of the substantive theory, which is fully grounded in the data (Glaser and Strauss, 1967; Strauss and Corbin, 1998). True to its qualitative research roots, this puts more emphasis on the quality of the data collected, so that the phenomenon can be better understood. In order to contextualise data collection and analysis, the use of the case study approach was found to be supportive of the principles of the grounded theory methodology, mainly the utilisation of multiple sources and constant comparison of the findings from these multiple sources. Each case study represents a 'cycle' of analysis, and, in a multiple case studies situation, the comparison involves further 'cycles' to be conducted, to the point of data saturation (Sutrisna and Barrett, 2007).

The other essential component of the research design, which is also a point of consideration here, is the literature review. Acknowledging the multiple perspectives of a diverse range of existing theories, it is important to discuss how they can be incorporated in designing the research methodology. Although keeping the connection with the actual data is crucial when developing the substantive theory in grounded theory, there is also a need to bring the existing theories into the development (Barrett and Stanley, 1999). The Glaserian approach to grounded theory does not recommend that researchers conduct a literature review prior to the data collection, whereas a preliminary literature review is 'acceptable' in the Straussian approach to grounded theory. The main purpose of the literature review in grounded theory is to support the development of the substantive theory – that is, mainly during data analysis, rather than as a prerequisite of data collection, as in deductive reasoning. Although there is an ongoing debate on the role of the literature review in grounded theory, which is also known as theoretical sensitivity, other scholars have seen this from a different perspective (Gunne-Jones, 2009; Setiawan, 2014): that an initial reading may actually improve theoretical sensitivity (McCreaddie and Payne, 2010). In the later development of the grounded theory methodology, scholars have suggested that the literature review should play an important role, particularly in the initial stage of a research (Heath and Cowley, 2004). These different views have serious implications for the design of research. Many researchers have found themselves in the situation where they have conducted a full-blown literature review (following the traditional or typical sequence of conducting research), just to find out that, by doing so, they may have excluded the grounded theory methodology as a potential approach. This further reinforces the need to fully understand research methodology and to design the research prior to performing other activities, even something as basic and 'harmless' as a literature review.

14.4 Data collection in grounded theory

When data are collected for research, one of the main considerations is sampling. In grounded theory, this has been described as theoretical sampling (Glaser and Strauss, 1967), which can be regarded as one of the non-probabilistic sampling techniques. However, different from the more generic, purposeful sampling commonly referred to in qualitative research (refer to Cutcliffe, 2000, for further discussion), theoretical sampling evolves during the data collection and analysis of grounded theory. Thus, decisions regarding further data to collect, further respondents to be interviewed, and so on, will be informed by the data gathered and analysed so far to develop the intended substantive theory (Strauss and Corbin, 1998). Even though the determination of the initial sampling would follow general purposeful sampling techniques, the subsequent sampling would be guided by the emergence of the substantive theory and what needs to be explored to further develop the substantive theory. Thus, in the further phase of data collection, theoretical relevance will help to direct the sampling process and guide the researchers in determining which respondents to interview from relevant groups in the field (Gunne-Jones, 2009). The subsequent process involving data collection, the coding process and analysis should later determine what follow-up questions should be asked in the generation of a substantive theory. This process will be repeated in such a way that the emerging theory 'controls' the process of data collection and analysis, to the point of saturation (Strauss, 1987).

When theoretical sampling is conducted, two factors should be considered (Glaser and Strauss, 1967). The first factor concerns the representativeness of the respondents within the field being researched. The second factor focuses on the potential to conduct theoretical comparisons within groups, or the representative of the groups; which groups should be chosen depends on two selection reasons: the ability to 'generate as many properties of categories as possible' and the ability to give 'accurate evidence for description and verification' (Glaser and Strauss, 1967: 49–50). Although it is acknowledged that the list of respondents to be interviewed will grow out of the theory emerging from the data analysis conducted so far, at the same time, it is also important to stay consistent within the research aim and objectives. This is particularly true for research in applied disciplines such as the built environment, which typically aims to solve more practical problems. This is not intended to limit the development of the substantive theory and restrict the researchers and respondents to include relevant matters, but to ensure the achievement of the research's original intention. Findings that are relevant but slightly beyond the scope of the research should be reported and recommended for further research.

Similar to many other interpretive methods, data collection in grounded theory is typically contextualised by case study. The next consideration would be the selection of case to be included and the respondents to be interviewed. Case study has been used to develop and test research hypotheses since the 1800s (Naumes and Naumes, 2006). Case study has been widely used and regarded as a strategy or an approach, rather than a research method (Robson, 2011). The capabilities, strengths and weaknesses of the case study strategy must be considered within a broad framework of conducting research in a particular topic (Sjoberg *et al.*, 1991). The selected research strategy typically involves a compromise between options, with the choices frequently determined by the availability of resources, as well as the nature of the problem itself (Gill and Johnson, 1997).

As a research strategy, the case study (which can be conducted as a single or multiple case study) can be perceived as an empirical enquiry conducted in the natural setting of the phenomenon being investigated (Yin, 2003). A multiple case study, which is the typical choice when case study research is designed, can enhance the potential to generalise the findings from the study (Naumes and Naumes, 2006). The use of multiple case studies is not intended to collate a 'sample' of cases to arrive at a generalisation regarding some population (as in statistical generalisation), but rather to have an 'analytic or theoretical generalisation' (Robson, 2011). Analytical or theoretical generalisation in this matter has been regarded as the utilisation of data collected in a particular study to provide theoretical understanding that contain a certain degree of universality, to allow their projection in other situations (Sim, 1998). The multiple case study strategy often aims to identify patterns using replication logic within and/or among cases, which can be either similar (literal replication) or contrary, but needed for predictable reasons (theoretical replication). Another strategy, known as pattern matching, compares findings across cases or with a theoretical proposition to identify patterns (Yin, 2003).

Following the grounded theory methodology, data collection and analysis should take place and only cease when data saturation or theoretical saturation is achieved. This can be considered one of the main common denominators of various views of grounded theory research. Saturation indicates the stability of the categories established so far, without 'new properties of the category emerg[ing] during data collection' (Charmaz, 2006, p. 12). This necessitates a researcher acquiring a sampling group that can provide a wide range of information (Gunne-Jones, 2009) to the point that another interview will not shed new light on anything that has been discussed in previous interviews. As the data analysis in grounded theory is conducted in parallel with further data collection, data saturation is not simply achieved by asking the respondents the same set of interview questions. Data saturation occurs from the data analysis, taking into account (relevant) new findings/directions that maybe revealed in, and triangulated through, satisfactory exploration of subsequent data collection and other sources, including (further) literature study, further evidence from archival analysis and so on.

14.5 Data analysis in grounded theory

As grounded theory requires interplay between analysis and data collection, coding for analysis is integrated with data-collection procedures. One way of establishing coding frames is to follow the sequence of open coding, axial coding and selective coding (Strauss and Corbin, 1998; Heath and Cowley, 2004). Open coding involves organisation of the data into categories and subcategories. In axial coding, the data are put back together in a new way through the establishment of relationships between these categories and subcategories (Blaikie, 2000). Selective coding involves conceptualisation of the relationships among categories for theory building (Groat and Wang, 2002). Selective coding also includes comparison of the potential theoretical frameworks emerging from analysis with the existing theories in the literature (Barrett and Stanley, 1999).

Table 14.1 Data-analysis procedures in grounded theory methodology

	Glaser	*Strauss and Corbin*
Initial coding	Substantive coding Data dependent	Open coding Use of analytic technique
Intermediate phase	Continuous with previous phase Comparisons, with focus on data, become more abstract, categories refitted, emerging frameworks	Axial coding Reduction and clustering of categories (paradigm model)
Final development	Theoretical (coding) Refitting and refinement of categories that integrate around emerging core	Selective coding Detailed development of categories, selection of core, integration of categories
Theory	Parsimony, scope and modifiability	Detailed and dense process fully described

Source: Heath and Cowley, 2004: 146

Following two main streams of the grounded theory methodology, Table 14.1. summarises the data-analysis procedures.

This following section discusses the use of coding strategies to interpret interview data, using the Strauss and Corbin terminologies to further discuss open coding, the process of abstraction, axial coding, selective coding (Heath and Cowley, 2004) and the use of a memo to document the ideas emerging during the analysis process (Gunne-Jones, 2009).

Open coding

Upon completion of each interview, the captured audio data will normally be transcribed into textual format for data-analysis purpose. So that the analysis can begin, the transcripts will need to be analysed thoroughly. They may be analysed at word-by-word level or at the level of textual chunks of data (Gunne-Jones, 2009) to develop a concept relevant to the research context (Strauss, 1987). This exercise is useful to generate sensitivity towards the topic, as well as ensuring that the research covers the emerging issues. Gradually, the researchers might feel more confident reading the text, from word by word to line by line, sentence by sentence, or even by paragraph. Although almost every single word in the initial analysis may have to be defined, in the later analysis process the researcher may be able to put a whole paragraph into a single node coding. In this initial phase, the researchers must be 'open' to the opportunity to use non-standardised terminology, which appears directly from the text, commonly called *in vivo* coding. The initial stage of coding usually involves a careful review of the text and its relevancy to the research context (McCreaddie and Payne, 2010). Hundreds of 'individual nodes' can well be developed at this stage, and the researcher is invited to create as many representative categories as possible (Glaser and Strauss, 1967). As a detailed analysis is conducted, some pattern might emerge (Gunne-Jones, 2009: 77). Although the emerging nodes may share particular similarities, characteristics or properties, the characteristics and properties of the emerging nodes would be best captured in the memo. This initial coding process provides a platform on which to develop a substantive theory.

Axial coding

Having developed categories in the form of nodes from the initial analysis phase (open coding), the researcher should start bringing together similar and relevant nodes and observing the emergence of natural groupings of the nodes. In this phase, axial coding emerges as the next step to 'develop higher level categories by speci-fying the condition, context, consequences' (Strauss, 1987; Gunne-Jones, 2009). The density of the explanation will increase by the time the analysis takes place. Progressively, not only are the codes being grouped and classified, but also the relationship among the groups of codes or nodes is being uncovered (Strauss, 1987). This process extends beyond the data generated in the initial process to a higher level of abstraction – but without the loss of any pertinent information or data richness (Strauss and Corbin, 1998; Barrett and Sutrisna, 2009) – in which the researchers become more focused (Charmaz, 2006) and analytical in coding the data. Based on the uncovered relationship and grouping into a hierarchical structure (known as tree nodes in NVivo terminology), a more holistic picture of the phenomenon/phenomena under investigation should start to emerge. To further complement this, the researchers will generally need to develop further discussion, explanation and/or definition to better understand the categories/nodes and build the connection between groups or categories. Again, this will typically involve the development of memos to capture their observation on the emergence of the categories.

Selective coding

As the researcher continues to merge nodes and organise them in a hierarchical manner to move up the level of abstraction, the final, top categories are known as the core categories. The core categories' selection has been considered to be 'systematically connected back to coding process' (Strauss, 1987: 33). It will select only the most relevant codes to develop the substantive theory. In other words, this step will focus on the most significant issues or elements within the phenom-enon/phenomena being researched. Less significant codes or elements will naturally appear as children nodes of these top-level categories as the main outcome of the analysis process. Development of the highest level of categories that represents the highest level of abstraction is known as selective coding. The substantive theory developed will typically be presented and structured under the core categories, as its 'backbone'. The very strong influence of this stage of analysis on the development of the substantive theory as the outcome of the research has put selective coding as the most important stage of the analysis in grounded theory. At this level of analysis, the researcher is also challenged to present overall understanding of the relationship between the core categories, in a holistic manner. The abstraction process requires the researcher to 'create a more dense integrated theory of greater scope' and 'collapse more empirical grounded categories into higher order conceptual ideas' (Goulding, 2002: 56–77). This process can be perceived as the bridge for the researcher to complete the transition from raw data to an insight or finding of the research.

Memo writing

The analysis phase typifies another important element that occurs during the analytical process, from open coding up to selective coding. A memo portrays any ideas coming out of the process of categorisation, abstraction and selection of coding. In the initial process of analysis, the researcher should commence the writing of the memo for each category developed. Although there are fewer 'restrictions' on developing memos for this early stage/lower level of categories (open coding), the later stages (axial coding and selective coding) require a more holistic view of the grouping and the taking into account of the relationship between categories. For instance, on an occasion where two earlier categories are combined to form a new category, a new memo for this new category should also be developed. This exercise provides an analytical 'playground' where the researcher can continuously challenge their observation. The continuous development of the memo – through addition and continuous rewriting following the axial and selective coding procedures – forms the basis of the substantive theory itself. As onerous as it may seem, memo writing helps to 'theoretically write up ideas, generate relationship, and direct future theoretical sampling' (Gunne-Jones, 2009: 65). Memos serve as a central element in the analysis process, which, in fact, becomes a 'compulsory' requirement in using grounded theory (Goulding, 2002). Research claiming to use the grounded theory methodology but failing to demonstrate the 'journey' of capturing salient issues and discussions through continuous development of the category memos to build the substantive theory will typically receive criticisms, and its soundness will be questioned.

14.6 The grounded theory 'style' of data analysis

This section is dedicated to discussing recent adaptations of the grounded theory methodology, including in built environment research, under the banner of 'grounded theory methodology style' data analysis. Thus, some researchers decided to use the data-analysis procedures and protocols of the grounded theory methodology, without being fully committed to implementing it as an overall research methodology. Questions typically (and understandably) arise from this around its credibility. Although one of the reasons behind this so-called 'partial adoption' of the grounded theory methodology may well be the phenomenon discussed earlier in this chapter, which involves researchers 'discovering' the grounded theory methodology only after conducting a full-blown literature review, in many cases, this 'partial adoption' was a deliberate choice, after careful consideration and based on a full understanding of research methodology design.

Strauss and Corbin (1998) clearly stated that the procedures were set mainly to help provide the process with standardisation and rigour, but were not designed to be dogmatic, constraining creativity and flexibility. Charmaz (2006) viewed grounded theory as a set of principles and practices, not as prescriptions or packages, and advocated the view that grounded theory can complement other approaches when qualitative data analysis is conducted, and mentioned examples of other researchers who acknowledged and implemented specific aspects of the approach. Thus, in some cases, grounded theory has not been viewed as a package for the conducting of research, and its components can be successfully applied in various research. In built environment research, there are examples where the grounded theory 'style' of data analysis has been adopted. In a UK-based research project, aiming to provide urban-development and urban-planning responses to social diversity and potential conflicts in Indonesia, Setiawan (2014) utilised grounded theory analysis, involving open coding, axial coding, selective coding and memo writing to analyse the interview data. The rigour of the grounded methodology in the data analysis has enabled the researcher to develop a framework at the personal level and community level, providing a developmental framework that can be used by government and urban-planning practitioners to improve the urban and rural environment, fostering social cohesion. In a Western Australia-based research aiming to identify the information gap between the construction phase and occupancy (facilities management) phase, Tan (2014) utilised the grounded theory methodology to analyse the interview data, which enabled the researcher to develop an information/data flow diagram to map all the relevant information and data exchanges necessary, before, during and after the handover to the client/user in commercial building projects. In these research projects, a thorough literature review was conducted prior to the data collection (Setiawan, 2014), and a preliminary literature review was carried out prior to the data collection, with an in-depth literature review carried out simultaneously with the data collection and analysis process (Tan, 2014).

In both cases, it is clear that the aim was not to develop a substantive theory, and the grounded theory analytical protocol has been used to support the development of a framework and information/data flow diagram. Strauss and Corbin (1998) acknowledged that theory building is not necessarily the goal of every research endeavour and welcomed researchers using the analytical techniques and procedures from grounded theory in different ways and blending them with the researchers' own techniques. Credibility of the findings in both research projects was demonstrated through external presentation and discussion of the resulting framework and information/data flow diagram with relevant practitioners to further refine and finalise them. Both research projects have benefited from the grounded theory analysis procedures and provided evidence of successful 'partial' adoption of grounded theory in the built environment field. However, researchers intending to 'partially' adopt grounded theory as an analytical tool should fully understand the complete methods to begin with and, only after achieving this holistic understanding, design and include a specific component of the grounded theory – that is, data analysis – to suit the needs of their particular research projects.

14.7 Summary and reflection

In the discussion about the original conception of the methodology, Goulding (2002) pointed out that the grounded theory aims to provide answers to research questions grounded in the data in an inductive manner. It moves forward from theory testing to theory building, and from description to conceptual categorisation. The grounded theory methodology has shown that qualitative research could also be conducted in a systematic and structured manner and, hence, can be considered credible through its 'distinctive procedures to aim at the level of abstract theorising' (Goulding, 2002: 36–7). This characteristic appeals to researchers, not only from pure social science, but also from developmental studies and many other disciplines, including the built environment. Research in architecture, for example, now recognises the use of grounded theory as a powerful research tool (Groat and Wang, 2002). In this regards, grounded theory emerges as 'an important research strategy in developing a substantive theory that aids understanding and actions in the area under investigation' (Heath and Cowley, 2004: 149).

Researchers in the built environment have successfully utilised the grounded theory methodology (e.g. Loosemore, 1999; Dainty *et al.*, 2000; Sutrisna and Barrett, 2007), mainly owing to its capability to draw abstraction and meaningful findings from complex data. Grounded theory has been used to provide means of arriving at research findings rather than being the end of the research itself. In previous built environment research, grounded theory has been used in triangulation with budget history analysis (Short *et al.*, 2007) as the precursor of cognitive mapping analysis (Barrett and Sutrisna, 2011) and as the basis of developing rich picture diagrams (Setiawan, 2014; Sutrisna and Barrett, 2007).

Despite its distinctive procedures, contemporary use of grounded theory recognises a more flexible utilisation of the methodology (Charmaz, 2006). Although the grounded theory methodology can be considered a holistic technique, evidence suggested that the nature of research sometimes opens up the need for the utilisation of a specific part of grounded theory, particularly the data analysis. In using this grounded theory 'style' of data analysis, researchers need to understand the main concept and the holistic mechanics of the grounded theory methodology before deciding to incorporate it in their research design.

From the ongoing discussion in this chapter, a number of important, key considerations can be summarised. First, a researcher should be aware of the influence of their philosophical stance on the design of their research methodology. This applies to all researchers, but more in the case for researchers adopting 'less traditional' tools and methods in built environment research, including grounded theory. The discussion of ontology and epistemology stances prior to research design will provide a better understanding of the methodological choices.

Second, as there are several directions/styles of grounded theory, a built environment researcher should fully familiarise him/herself with each direction before committing their research towards a particular direction/style. Inductive process is the common denominator of these directions/styles and should be the core approach in analysing the data in grounded theory research (Heath and Cowley, 2004).

Third, despite the divergence of views about literature reviews in the discussion of the grounded theory methodology, the more contemporary views advocate a literature review as an essential component in developing sensitivity towards the theoretical context of the emerging findings from the data. However, this theoretical sensitivity should not constrain the emergence of new categories from the interviews. This is an opportunity for the researcher to better understand the context of the area being researched, so that they can ask more effective questions in the interviews and capture richer data, allowing the discussion to develop into other relevant issues but, at the same time, keeping the discourse within the research parameters.

Fourth, the grounded theory methodology claims the use of memo writing as a key characteristic (Goulding, 2002). Therefore, when grounded theory is used, it is highly recommended that memos are developed and continuously updated in light of new information. The memos will form the basis of the substantive theory itself, but it can be considered good practice to develop memos, even when the purpose of the research is not to develop a substantive theory.

Finally, it is important to anticipate the complexity of potential emerging data and the number of participants sufficient to produce data saturation, to allow the process of constant comparison and theoretical sampling (Goulding, 2002; Onwuegbuzie and Collins, 2007).

References

Amaratunga, D., Baldry, D., Sarshar, M. and Newton, R. (2002). 'Quantitative and qualitative research in the built environment: Application of "mixed" research approach', *Journal of Work Study*, 51(1): 117–31.

Barrett, P. and Stanley, C. (1999). *Better Construction Briefing*. Oxford, UK: Blackwell Science.

Barrett, P. and Barrett, L. C. (2003). 'Research as a kaleidoscope on practice', *Construction Management & Economics*, 21(7): 755–66.

Barrett, P. and Sutrisna, M. (2009). 'Methodological strategies to gain insights into informality and emergence in construction project case studies', *Construction Management & Economics*, 27(10): 935–48.

Barrett, P. and Sutrisna, M. (2011). The internal dynamics of the projects. In C. A. Short, P. Barrett and A. Fair (eds), *Geometry and Atmosphere: Theatre buildings from vision to reality* (pp. 223–8). Farnham, UK: Ashgate.

Blaikie, N. (2000). *Designing Social Research* (1st edn). Cambridge, UK: Polity Press.

Charmaz, K. (1990). 'Discovering chronic illness: Using grounded theory', *Social Science & Medicine*, 30(11): 1161–72.

Charmaz, K. (2000). Grounded theory: Objectivist and constructivist methods. In N. Denzin and Y. Lincoln (eds), *Handbook of Qualitative Research* (2nd edn, pp. 509–35). Thousand Oaks, CA: Sage.

Charmaz, K. (2006). *Constructing Grounded Theory: A practical guide through qualitative analysis*. London: Sage.

Chynoweth, P. (2013). 'Practice-informed research: An alternative paradigm for scholastic enquiry in the built environment', *Property Management*, 31(5): 435–52.

Crotty, M. (1998). *The Foundations of Social Research: Meaning and perspective in research process*. London: Sage.

Cutcliffe, J. R. (2000). 'Methodological issues in grounded theory', *Journal of Advanced Nursing*, 31: 1476–84.

Dainty, A. R. J. (2008). Methodological pluralism in construction management research. In A. Knight and L. Ruddock (eds), *Advanced Research Methods in the Built Environment* (pp. 1–13). Chichester, UK: Wiley-Blackwell.

Dainty, A. R. J., Bagilhole, B. M. and Neale, R. H. (2000). 'A grounded theory of women's career under-achievement in large UK construction companies', *Construction Management & Economics*, 18: 239–50.

Del Casino, V. J., Grimes, A. J., Hanna, S. P. and Jones III, J. P. (2000). 'Methodological frameworks for the geography of organizations', *Geoforum*, 31: 523–38.

Denzin, N. K. (1993). Sexuality and gender: An interactionist/poststructural reading. In P. England (ed.), *Theory on Gender/Feminism on Theory* (pp. 199–222). New York: Aldine de Gruyter.

Gann, D. (2001). 'Putting academic ideas into practice: Technological progress and the absorptive capacity of construction organizations', *Construction Management & Economics*, 19: 321–30.

Gill, J. and Johnson, P. (1997). *Research Methods for Managers* (2nd edn). London: Sage.

Glaser, B. G. (1978). *Theoretical Sensitivity: Advances in the methodology of grounded theory*. Mill Valley, CA: Sociology Press.

Glaser, B. G. (1999). 'The future of grounded theory', *Qualitative Health Research*, 9(6): 836–46.

Glaser, B. G. and Strauss, A. L. (1967). *The Discovery of Grounded Theory: Strategies for qualitative research*. Chicago, IL: Aldine Press.

Goulding, C. (2002). *Grounded Theory: A practical guide for management, business and market researchers*. London: Sage.

Griffiths, R. (2004). 'Knowledge production and the research-teaching nexus: The case of the built environment disciplines', *Studies in Higher Education*, 29(6): 709–26.

Groat, L. and Wang, D. (2002). *Architectural Research Methods*. New York: Wiley.

Gunne-Jones, A. (2009). *Town Planning 2009: A practical guide, the essential guide for today's planning system*. Coventry, UK: RICS Books.

Heath, H. and Cowley, S. (2004). 'Developing a grounded theory approach: A comparison of Glaser and Strauss', *International Journal of Nursing Studies*, 41: 141–50.

Hunter, K. and Kelly, J. (2008). Grounded theory. In A. Knight and L. Ruddock (eds), *Advanced Research Methods in the Built Environment* (pp. 86–98). Chichester, UK: Wiley-Blackwell.

Lomborg, K. and Kirkevold, M. (2003). 'Truth and validity in grounded theory – a reconsidered realist interpretation of the criteria: Fit, work, relevance and modifiability', *Nursing Philosophy*, 4(3): 189–200.

Loosemore, M. (1999). 'A grounded theory of construction crisis management', *Construction Management & Economics*, 17(1): 9–19.

McCreaddie, M. and Payne, S. (2010). 'Evolving grounded theory methodology: Towards a discursive approach', *International Journal of Nursing Studies*, 47: 781–93.

Mingers, J. (2004). Re-establishing the real: Critical realism and information systems. In J. Mingers and L. P. Willcocks (eds), *Social Theory and Philosophy for Information Systems Research* (pp. 372–406). Chichester, UK: John Wiley.

Naumes, W. and Naumes, M. J. (2006). *The Art and Craft of Case Writing*. London: Sage.

Onwuegbuzie, A. J. and Collins, K. M. T. (2007). 'A typology of mixed methods sampling designs in social science research', *The Qualitative Report*, 12: 281–316.

Raftery, J., McGeorge, D. and Walters, M. (1997). 'Breaking up methodological monopolies: A multi-paradigm approach to construction management research', *Construction Management & Economics*, 15(3): 291–7.

Robson, C. (2011). *Real World Research* (3rd edn). Chichester, UK: John Wiley.

Rooke, J., Seymour, D. and Crook, D. (1997). 'Preserving methodological consistency: A reply to Raftery, McGeorge and Walters', *Construction Management & Economics*, 15(5): 491–4.

Runeson, G. (1997). 'The role of theory in construction management research: Comment', *Construction Management & Economics*, 15(3): 299–302.

Ryan, G. W. and Bernard, H. R. (2003). Data management and analysis methods. In N. K. Denzin and Y. S. Lincoln (eds), *Collecting and Interpreting Qualitative Materials* (2nd edn, pp. 259–309). Thousand Oaks, CA: Sage.

Schwandt, T. A. (1998). Constructivist, interpretivist approaches to human inquiry. In N. K. Denzin and Y. S. Lincoln (eds), *The Landscape of Qualitative Research: Theories and issues* (pp. 221–59). London: Sage.

Setiawan, W. (2014). 'Urban development and the urban planning responses to social diversity and potential conflict in Indonesia'. Unpublished PhD thesis, School of Built Environment, University of Salford, UK.

Seymour, D., Crook, D. and Rooke, J. (1997). 'The role of theory in construction management: A call for debate', *Construction Management & Economics*, 15(1): 117–19.

Short, C. A., Barrett, P., Dye, A. and Sutrisna, M. (2007). 'Impacts of value engineering on five capital arts projects', *Building Research & Information*, 35(3): 287–315.

Sim, J. (1998). 'Collecting and analysing qualitative data: Issues raised by the focus group', *Journal of Advanced Nursing*, 28: 345–52.

Sjoberg, G., Williams, N., Vaughan, T. R. and Sjoberg, A. F. (1991). The case study approach in social research. In J. Feagin, A. M. Orum and G. Sjoberg (eds), *A Case for the Case Study* (pp. 27–79). Chapel Hill, NC: University of North Carolina Press.

Stern, P. N. (1980). 'Grounded theory methodology: Its uses and processes', *Image*, 12: 20–3.

Strauss, A. L. (1987). *Qualitative Analysis for Social Scientist* (14th printing 2003). Cambridge, UK: Cambridge University Press.

Strauss, A. L. and Corbin, J. (1998). *Basics of Qualitative Research: Techniques and procedures for developing grounded theory* (2nd edn). Thousand Oaks, CA: Sage.

Sutrisna, M. (2009). Research Methodology in Doctoral Research: Understanding the meaning of conducting qualitative research. Working Paper, *ARCOM Doctoral Workshop*, Liverpool, UK, 12 May.

Sutrisna, M. and Barrett, P. (2007). 'Applying rich picture diagrams to model case studies of construction projects', *Engineering, Construction & Architectural Management*, 14(2): 164–79.

Tan, A. Z. T. (2014). 'Building handover versus operations and maintenance: An investigation of the information gap'. Unpublished dissertation, Department of Construction Management, Curtin University, Western Australia.

Yin, R. K. (2003). *Case Study Research: Design and methods* (3rd edn). London: Sage.

15 Grounded theory style analysis in action

Utilising multi-layer methods in developing built environment response to social conflicts in Indonesia

Wisnu Setiawan, Monty Sutrisna and Peter Barrett

A number of researchers have attempted to explain why and how social conflicts have occurred in Indonesia over the history of the nation. Both research and response to the conflicts have focused mainly on rather short-term dispute settlement. Researchers have also viewed the incidents from various perspectives, from sociopolitical to economic. However, current research cannot agree why social conflicts actually happened, particularly owing to the diverse use of methodology and the variability of the conflicts. Research into this topic, therefore, is increasingly using a case-specific approach or one that involves several cases with similar characteristics. Research into social conflicts has traditionally utilised interpretive and inductive approaches to bring both the individuality of the actors and contextual aspects of the situations into their investigation. In order to bring about specific, yet comprehensive, understanding of conflicts, this research implemented a multiple case studies approach, with in-depth interviews as the data collection technique. For analysis of the complex, rich data, the grounded theory methodology style has been selected as the most appropriate method. This chapter discusses the design and implementation of the grounded theory methodology style in analysing the data. The discussion is particularly focused on two key elements: the position of the literature review and the grounded theory methodology as a method. This research has successfully adopted the grounded theory methodology style of analysis to reveal that social conflicts have a reciprocal relationship with the practice of urban development in the past and present. It also suggests that a 'good' built environment could facilitate a reduction in the potential for conflict in the future.

15.1 Background

Most research on conflict related to urban development has focused mostly on developed countries (e.g. Bollens, 2006; Murtagh *et al.*, 2008). Research could extend to a number of countries and cities, but, in the context of developing countries, accessible resources and literatures lead to one specific place: Indonesia. The history of violent communal conflict, in a contemporary context, has emphasised Indonesia as being unique, particularly after its crisis in 1998. Indonesia is considered 'extraordinarily diverse' (Mancini, 2008: 115), with a relatively low ethnic homogeneity index (45 per cent) and a low gross national product ($1,100 per capita; Kurian, 2001). Ethnic homogeneity often reflects 'national strength', and per capita income could be the most used indicator of economic development (Kurian, 2001). Following the more recent economic crisis at the end of the 1990s, the country experienced increased interregional inequality – for example, the decrease in the employment rate, particularly in the urban setting (Kataoka, 2010). A research found that the characteristics of economic growth, poverty, human development and education have contributed to the occurrence of conflicts (Tadjoeddin and Murshed, 2007). The economic crisis was followed by the emergence of all types of conflict, from national to local, political to community, by state or by community (Colombijn and Lindblad, 2002; Dijk, 2002).

Research on violent conflict in Indonesia has focused on various perspectives that have mainly related to racially driven emotion, economic frustration, urban poverty, racial/ethnic relationships, political matters, religious differences and so on (e.g. Drexler, 2006; Loveband and Young, 2006; Poerwanto, 2006; Cahyono *et al.*, 2008). In general, current research shows that the country experienced various types of violence in particular spatial and time contexts, which implies that observations on the 'relationship' among societies became an important issue to take into account (Bollens, 2006; Purdey, 2006) in the context of urban development. In addition to that, scholars seem unable to reach a consensus on whether social diversity and spatial contexts could be directly linked to conflicts. Although recent violent conflicts have transformed the pattern of current development (Koeswinarno *et al.*, 2004), for example in relationship to forestry and land management, the literature presents little discussion on how violent conflicts have reciprocally been linked to urban policy and development.

It appears that research into violent conflict in Indonesia challenges the use of an 'appropriate' method to reveal the full story. The challenge of explaining the violent conflict in Indonesia acknowledges a different scope, exclusion and level of analysis (Bertrand, 2008). Therefore, research is growing, especially that using the case study method (Barron *et al.*, 2009). By using case study, the research is expected to be able to reveal and explain potential factors and their relationship (Vaus, 2001), regarding which particular occurrence affects which particular event (Hancock and Algozzine, 2006) or even the causal process of an issue (Hammersley and Gomm, 2000; Vaus, 2001; Yin, 2009). It indicates that case study research is concerned with the 'why' and 'how' of the research question. Thus, it meets the research questions of the study: why the cultural gap and social diversity remain and affect the potential conflict; how the role of urban development responds to or manipulates this diversity. This research also proposes that understanding of the phenomenon is best approached through observation of a number of cases with similar characteristics, as comparative study within the same group would poten-tially result in a better understanding of the conflict (Bertrand, 2008). In summary, this research employs multiple case studies in three different locations in Indonesia, representing the social diversity that led to social conflict in the cases. A holistic approach focuses on the practice of urban planning and development in response to social diversity.

The research uses case study methods in order to better understand the situation. In order to better understand the phenomenon, research could rely on the use of interviews, particularly guided, in-depth interviews (Charmaz, 2006), so that it can explore the richness of the issues (as discussed in the previous chapter). By adopting this strategy, this research has generated meaningful and rich textual data. Following careful consideration of the data-analysis process and techniques available, the use of the grounded theory methodology (GTM) was considered a robust data-analysis technique for this research. It has been argued that the utilisation of a 'pure' GTM (please refer to Chapter 14 in this book) is not always the best option or practicable. This chapter presents the successful adoption of 'grounded theory analysis style' as a stand-alone data analysis technique.

15.2 Research methodology

Research philosophy

'Research philosophy contains important assumptions about the way in which the researchers view the world' (Saunders *et al.*, 2009: 108). Discussion about research methodology often begins with three components: ontology, epistemology and axiology. Using the construction of a house as an analogy, a research philosophy consisting of 'ontology and epistemology can be considered as the foundation upon which research is built' (Grix, 2010: 57). Understanding of the research philosophy helps in approaching the issue of diversity and the potential for conflict in an urban area of a complex nature. The relationship among the factors is often unclear. Some urban environments in developing countries exist with a high degree of inequality, without or with less violent conflict. On the other hand, developed countries with the notion of a higher degree of equality still experience violent conflict. The issue of diversity seems to be temporal, as it is situational. Not only do the tangible factors affect it, but also the intangible elements, such as perception. A mental image of the diversity differs from people to people, and from culture to culture. Research on this topic should ideally include human interaction as an important factor. The ontological stance of this research can be considered closer to constructivism. A constructivist position implies that reality – social phenomena, for example – is a product of social interaction and is continuously refined (Sutrisna, 2009; Grix, 2010) as it is subjective (Creswell, 1994; Saunders *et al.*, 2009).

Epistemology, on the other hand, is about 'how we come to know what we know' about reality (Grix, 2010: 63). It focuses on the process of constructing knowledge, the way to understand knowledge, the development of knowledge (Grix, 2010) and what represents knowledge (Saunders *et al.*, 2009). This research aligns with the interpretive stance. This particular epistemological position seeks to understand the differences created by the perceived view of reality. Research in this stance typically interprets reality from stories or informal and unstructured interviews.

Axiology, the third common component of a research methodology discussion, relates to the assessment of value (Saunders *et al.*, 2009). A discussion about axiology links to the approach a researcher uses to understand the study. A discussion about axiology links to the approach of a researcher in understanding the study. In axiology, two contrasting stances to be declared concern the positioning of this particular research, whether it is value-dependent or independent of the researcher's influence (Saunders *et al.*, 2009). This research exemplifies the axiological stance that personal background has a role in shaping the study. It offers the chance to express a personal statement relating to the topic.

Research strategy

As this research looks at a specific, complex phenomenon, it does not attempt to find large-scale generalisation, although application of the findings to specific circumstances similar to the cases being studied is also possible. When a research deals with a number of the population to create generalisation, it tends to employ larger numbers of participants or objects. This research chose to focus on a smaller, unique phenomenon with a deeper understanding.

In order to offer understanding of the nature of research problems, this study attempts to build an explanation of 'why it happens' and 'how to respond'. It aims to explain the relationship between various factors and, later on, to apply the findings to similar types of problem. The nature of such research allows it to correlate and explain various factors' relationships (Hancock and Algozzine, 2006; Saunders *et al.*, 2009). Therefore, this study favours explanatory research to explain the relationship between a particular phenomenon and specific factors in an attempt to respond to the question: 'why is it going on?' (Vaus, 2001: 1). To provide an in-depth, comprehensive understanding of the issue, a significant part of this research follows the inductive approach, using the data collected from the field. This approach is applicable particularly to understanding the role of human interaction in particular incidents (Saunders *et al.*, 2009). The use of qualitative data with a more flexible structure allows the researcher to get involved with the object being researched (Sutrisna and Barrett, 2007; Sutrisna, 2009).

The strategy to focus on fewer cases points to an argument that comparative study performs well in case study research (Hammersley and Gomm, 2000). This is also in line with the popular use of the case study method to conduct research into social conflicts (Barron *et al.*, 2009). The multiple case approach, in particular, has the possibility to capture various situations, or even a contrasting situation, in which the hypothesis will then need to consider potential differences between the cases (Yin, 2009). A multiple case study generally has to deal with more diverse information and relatively more numerous features of each case (Hammersley and Gomm, 2000). Figure 15.1 illustrates the case study approach in this research.

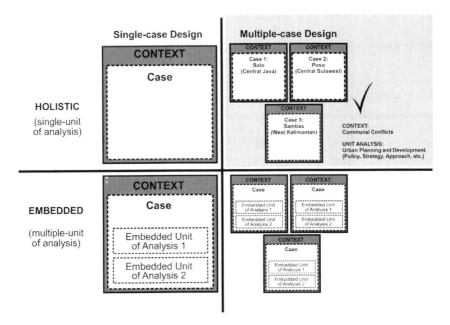

Figure 15.1 Position of the research based on case study typology
Source: Modified from Yin, 2009: 46

Case studies can begin with data collection and analysis to build a theory or model using categorisation (Strauss, 1987). The model that emerges from the initial phase of research should present clear 'theoretical elements and their connections with each other' (Strauss, 1987: 220). The last phase requires careful selection of the data that can support the emerging concepts or theory. This argument has brought about the use of the GTM to extract a more theoretical proposition from the data.

Data-collection strategy

In the context of Indonesia, research into violent conflicts using quantitative data appears to be able to provide a broad picture of the incidents across the country (Panggabean, 2006). However, these studies have some limitations. For example, McLaughlin and Ari (2010) present the linkage between the local dynamics of conflicts, dispute resolution, the socio-economic element and governance factors across twenty-nine provinces, but they only focused on 2 years of data and three main sets of questions. Similarly, a study of conflicts employing available data provided by IFL and UNSFIR (Muller and Vothknecht, 2011) covers about 60 per cent of all the incidents, but only in six provinces, focusing mainly on Java, West Nusa Tenggara and South Sulawesi. The study focuses only on some areas with little destructive impact and few fatalities. The use of quantitative data in studying violent conflicts in Indonesia indicates that the emphasis at the local level is limited (Bertrand, 2008). An approach at the local level appears potentially to provide a greater in-depth understanding of the dynamics at play (Panggabean, 2006). In short, the available statistical data have been considered insufficient for the study of conflicts, and collecting quantitative data from primary resources in the first place may pose yet another challenge.

Hence, this research focuses on the qualitative approach to data collection. This approach enables the research to gain deep understanding, while keeping the possibility of more information wide open. This idea, in particular, coincides with the characteristics of using a guided, open-ended interview style to ensure flexibility in 'sequencing and wording questions in different situations', without losing the detail but allowing deep information to emerge (Hancock and Algozzine, 2006: 3). A set of triggered questions is developed to provide a general structure to the interview:

- Triggered question 1: How was the situation before the conflicts?
- Triggered question 2: How were the conflicts emerging?
- Triggered question 3: How did the stakeholders react to this situation?
- Triggered question 4: What is the role of urban development?
- Triggered question 5: How should urban policy respond?

This research invited participants from four different stakeholder groups related to urban planning and development practices. They came from different backgrounds, such as government officers, particularly related to the urban planning authority, expert, researcher, urban planner and a local leader as a representative of sociocultural groups within the society. The interviews took place in more casual conversations to encourage the richness of the information. In order to ensure the conversation would not go beyond the topic, the use of literatures could evolve into a passive background that could help indirectly guide the data collection within the scope of the study (Table 15.1), 'enhancing sensitivity' to the discussion (Strauss and Corbin, 1998: 50). Keeping those features in mind, the research refers to the objectives and its relation with the expected information emerges from the data.

Data-analysis strategy: the use of grounded theory style

Along with the use of case studies, the analytical approach in much qualitative research, at some point, can somehow be considered to follow grounded theory style (GTS) for text analysis (Strauss, 1987: 218). The use of a case studies approach in the first place opens up the possibility to build theory or model using categorisation (Strauss, 1987), which corresponds to GTM as an analytical instrument: a method. Grounded theory shows its use as an analysis 'tool rather than recipes to follow' (Charmaz, 2006: 10): 'methods are merely tools' (Charmaz, 2006: 15). It invites researchers to choose the appropriate method of analysis that suits the nature of the research question. The initiative to involve GTS as an analysis

Table 15.1 Key features on urban planning and development in response to communal conflicts

Features	Issues
Political	Power disparity: – Governance – Multiple players – Civic culture
Economic	Inequality: – Economic development – Resources – Power balance
Cultural	Group identity: – Behaviour – Norm and value – Symbol
Physical	Segregation: – Land and property management – Urban settlement – Infrastructure provision

tool is related to a claim that, 'Grounded Theory provides systematic and flexible guidelines for collecting and analysing qualitative data to construct theories grounded in the data themselves' (Charmaz, 2006: 16). Grounded theory allows research into 'interaction, action, and processes' (McCreaddie and Payne, 2010: 781).

Grounded theory initially emerged with the influence of symbolic interactionism to merge the social and psychological concept (Goulding, 2002). Psychological research usually focuses on social behaviour, using general terms borrowed from a logical process. Sociological research looks more at an individual level that is determined by the social environment. The idea of social interaction puts both issues together in one understanding: how the social process affects personal behaviour, and vice versa. Glaser and Strauss's initial idea on grounded theory emerged in response to the fact that qualitative research at that time could not provide enough 'scientific' evidence (Glaser and Strauss, 1967). Qualitative research happened to be 'biased and anecdotal' (Charmaz, 2006: 5). Qualitative researchers were losing their credibility in front of quantitative researchers. Glaser and Strauss (1967) formulated a 'procedure to collect and analyse qualitative data, which is grounded in the behaviour, words, and actions' (Goulding, 2002: 40). Grounded theory tried to figure out a particular 'logic to construct abstract theoretical explanation' and to 'generate theory' (Charmaz, 2006: 5). As a result, more researchers now accept that grounded theory's procedures have become acceptable, rigorous and useful.

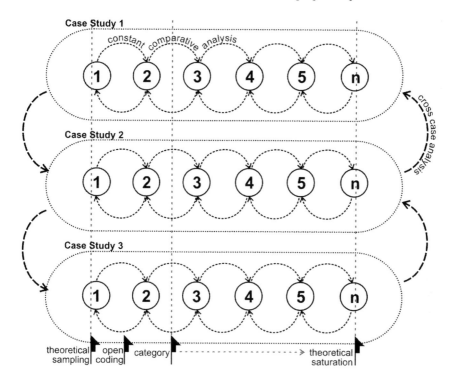

Figure 15.2 Illustrative analysis process of grounded theory style

The initial idea of grounded theory has been developing into at least two main streams. The first stream follows the 'Glaserian' style, and the other one follows the 'Strauss and Corbin' style (Goulding, 2002). Apart from the development of grounded theory into different styles, four key features appear in common: (1) theoretical sensitivity, (2) theoretical sampling, (3) constant comparative analysis and (4) theoretical saturation (Charmaz, 2006; McCreaddie and Payne, 2010). Figure 15.2 illustrates the general process of the grounded theory style adopted in this study.

Theoretical sensitivity: the position of literatures

This characteristic has appeared as one of the main debates in grounded theory research. The general understanding presumes that the GTM encourages a researcher to enter the field with limited information about the research object. However, the later development of GTM has invited the literature review to take an important role in the initial stages of a research (Heath and Cowley, 2004). This research acknowledges the existence of substantial initial reading to improve theoretical sensitivity (McCreaddie and Payne, 2010). This theoretical relevance will help to direct the sampling process, which represents groups existing within the field (Goulding, 2002).

The process: Theoretical sampling, comparison analysis and theoretical saturation

In the next phase, the research process involves data collection, coding and analysis. It later instructs what data should be gathered next. The process then will somehow generate a theory from the chosen sampling. This process will go on and on to direct the next sampling process. Ideally, the emerging theory controls the process of data collection and analysis (Strauss, 1987). During theoretical sampling, two key features should be considered (Glaser and Strauss, 1967: 46): the representation of groups within the field being researched and the potential for doing theoretical comparison within groups, or the representative of the groups. It should be able to 'generate as many properties of categories as possible' and could give 'accurate evidence for description and verification' (Glaser and Strauss, 1967: 49–50). This research anticipated those two features by inviting a 'sufficient' number of participants from four key stakeholders related to social conflict and the built environment.

The analysis of the interview employs constant comparative analysis across three cases to look for emerging 'patterns and themes' (Goulding, 2002: 69). One way of doing it is by looking at similarities and differences to generate classification and its attributes. The emerging issues from this comparison help to illustrate a concept (Glaser and Strauss, 1967), one that is relevant to the phenomenon, but not necessarily representing the 'whole field'.

This continuous process runs until the analysis becomes 'saturated'. Theoretical saturation indicates that, 'no new properties of the category emerge during data collection' (Charmaz, 2006: 12). It requires a researcher to acquire a sampling group that can provide a wide range of information (Goulding, 2002).

Methods in use: A summary

In summary, this research employs the case study approach to contextualise the scope of the study. This study focuses on three cases that experienced conflicts at a communal level during the same period of time, but in different geographical areas. In order to collect the information from the study field, forty-two participants in total were interviewed during the main data-collection phase, and six more participants were interviewed during the verification phase.

In the analysis of the data, this research uses the approach developed from the GTM. It begins with the first batch of six interviewees to generate the emerging initial categories through 'open coding'. At this stage, the research expects some '*in vivo* coding' represented by the emerging 'free notes'. In the next stage, it helps to build axial coding, the relationship, across the categories, as well as across the three cases. The last stage presents some important, key points that represent all the issues emerging from the data. In short, the research has brought together the case study approach and the GTS analysis, which are graphically presented in Figure 15.3. Table 15.2 summarises the methodological discussion, which begins with definition of aims and objectives at the philosophical level and proceeds to the application of research methods.

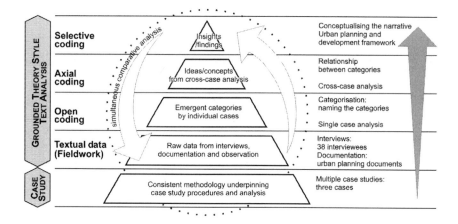

Figure 15.3 The use of case study and modified grounded theory style

Source: Developed from Dijk, 2002; Barrett and Barrett, 2003; Coppel, 2006

Table 15.2 The research methodology developed for this research

Aims	To develop a framework of urban planning and development direction in response to social diversity and conflicts (framework)
Epistemology	Human interaction is an important key to urban development (interpretive)
Ontology	The role of urban development to respond to diversity issues is perception-based (constructivism)
Axiology	'Ideas' about the role of urban development in the issues of conflicts (value-laden)
Research methodology	Research will employ 'continuous comparison' to confirm the understanding of the issue ('triangulation')
Methods	Time and space context is important for deep understanding of the issues of social diversity (case study; GTS)
Sources	Qualitative: interview, archives (qualitative)

Source: Modified from Grix, 2010

15.3 Findings and discussion

Case studies

In order to better understand the nature of the conflicts, this research has picked three cases: (1) Solo (Central Java), (2) Poso (Central Sulawesi) and (3) Sambas (West Kalimantan). The three cases illustrate that the communal conflict in Indonesia emerged with different motives. Socio-economic factors appeared to be one motive in Solo. The conflict has been associated with the gap between the Chinese ethnic group, which was alleged to control economic activity, and the Javanese, the indigenous people who suffered from the unequal development (e.g. Purdey, 2006). Socio-economic motives might also present as one reason in Sambas, where the people have to struggle for work according to ethnicity and occupational pattern (e.g. Arafat, 1998; Koeswinarno and Abdurahman, 2006; Peluso, 2008). However, the latest conflict showed that it was not only socio-economic motives but also cultural reasons that drove the Malay people to force the Madura people to leave their property. Cultural issues could be problematic in West Kalimantan, as they appear to be problematic in Poso too. The communities there used to live peacefully side by side, respecting the differences of culture and religion (Abdulrahman *et al.*, 2003). The conflict, however, has blurred the distinction between culture and religion. It involved people from different ethnic groups, coincidently with different religious backgrounds – Muslim and Christian.

The government has attempted to deal with the issues of conflict. Short-term efforts have attempted to resolve the conflicts under different labels. For example, some meetings provided local leaders from different sociocultural groups with the opportunity to sit together during the reconciliation process. In order to explore the opportunity of a long-term programme, the research has qualitatively approached a series of interviews as primary data. The research involved a number of participants meeting the general requirements of the research methods, which provided sufficient data for theoretical saturation.

The research identified and recruited the interviewees using the snowball approach. This technique particularly works in this research as it deals with quite sensitive issues. The research had to get information from a 'gatekeeper', who then led to other participants. The interviewees come from four main stakeholders, namely a government body that deals with urban planning, a researcher, an urban planner and a local leader representing a sociocultural group.

Analysis: Grounded theory style

Following the GTS to analyse interviews, the transcript analysis started with an initial theoretical sampling to find guidance for the next analysis steps. In this initial stage, the analysis attempts to capture interviewees' *in vivo* expressions. The emerging categories started with 'freestanding issues' at the beginning and captured 398 'free nodes'. The next phase attempted to see the pattern of those nodes, based on their similarity, differences and uniqueness. Eventually, it produced 233 subcategories, which were then grouped in thirteen categories. This included issues regarding culture, interaction, conflicts, natural resources, economic development,

Table 15.3 Summary of number of participants in relation to sampling guidance

Method	Sampling guidance	Number of participants
Case study	3–5 participants	3 cases: Solo, Poso and Sambas
Interview	12 participants	Solo: 13; Poso: 13; Sambas: 16
GTS	15–20 or 20–30 participants	42 participants (in total)

Source: Summarised from Onwuegbuzie and Collins, 2007: 288–9

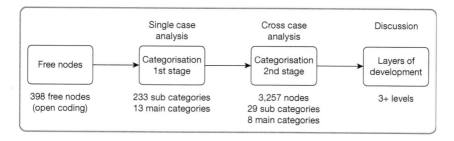

Figure 15.4 Category evolution of emerging issues

politics, segregation, people movement, infrastructure, social development, urban planning, spatial management and urban policy implementation in general. This first stage of categorisation provided a platform for the next stage of coding. It took up more coding from the interview, which led to the second stage of categorisation.

The final stage of the analysis recorded 3,257 references in total, which represent the number of 'quotations' or 'arguments' that appeared in the interview. In NVivo terminology, 'source' represents the number of participants who mentioned an issue. Although the research does not attempt to focus on quantification of the data, Table 15.4 clearly shows that the respondents most likely mentioned the cultural element and the issue of development when they talked about the recent social conflicts. NVivo has captured these emerging issues into a hierarchy of nodes, examples of which are presented in Figure 15.5 (cultural elements) and Figure 15.6 (urban planning and development elements).

At this stage, the nodes achieved the level of saturation and formed eight categories, with twenty-nine subcategories. These 'nodes' have slowly taken shape into a more manageable number of categories, followed by narrative. Across these eight main nodes, the pattern has shown a sense of levelling that spans all three cases. The levelling groups the findings in three layers of development: personal, community and urban. In other words, the interviewees noted the relationship between communal conflicts and the performance of urban development. In most cases, the role of social activities has been considered key to support a socially cohesive community. Figure 15.7 illustrates how the categories emerging in the final stage comprise up to eight main categories and three layers of development, with two additional elements of development in the context of social conflict.

In general, the interviewees from all the cases indicated that communal conflicts in some cases began with personal frictions, which may be linked to antisocial behaviour. At the community level, communication at the neighbourhood level and social infrastructures were perceived to be crucial. This implies that the conflicts also have a notion of sociocultural differences, which are believed to have derived from ethnic or religious beliefs. The findings also found an association between the recent conflicts and urban-level development. The emerging issues

Table 15.4 Summary of number of emerging nodes

References	Source	Main categories
254	33	1: Recent communal conflicts
120	31	2: Implication and precedent
110	26	3: Personal level
871	37	4: Cultural elements
266	33	5: Community-level development
384	35	6: Economic development
937	37	7: Urban planning and development
315	35	8: Governance

3,257

2-Tree Nodes		
Name	Sources	References
4-Cultural elements	37	871
⊟ Cultural interaction	36	477
Cultural and social event and activity	26	60
⊞ Diversity and cohesity	25	73
⊞ Interaction	36	344
⊟ Cultural transformations	13	23
Cultural change	11	19
⊞ Ethnicity transfer	3	4
⊟ Social group identification	36	323
⊞ Cultural definition	18	79
⊞ Ethnicity	21	73
Religion role	19	56
Stereotype	10	21
⊞ Symbols	26	94
Traditional value or norm and social control	16	40

Figure 15.5 Example of emerging nodes: Cultural elements

2-Tree Nodes		
Name	Sources	References
7-Urban planning	37	937
⊞ Concepts and implementations	24	183
⊞ Focus themes	32	146
⊞ Infrastructures	33	220
⊞ Regulation	19	60
⊞ Spatial management	31	154
⊞ Urban design	19	70
⊞ Urban planning document	17	71
⊞ Urban planning elements	15	28

Figure 15.6 Example of emerging nodes: Urban planning and development elements

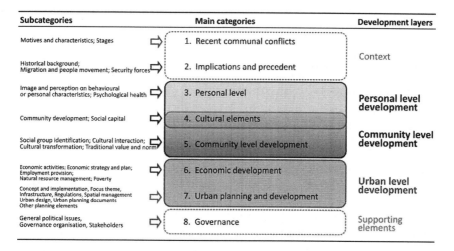

Figure 15.7 The emergence of main categories and development layers

touch upon matters of economic development strategy, the role of infrastructure at regional level, interregional development strategy and so on.

The layers of categories appear to comprehensively describe the complex situation related to communal conflicts and urban development. The layers help to guide the development of an urban-planning framework, in the next research development, which links to three different stakeholders in the development process – people, government and planning practitioners – as well as other stakeholders.

15.4 Conclusion

In the initial stage, research has to deal with the interrelationship between the research questions and the choice of research methodology. This research began with the question, 'Why does social conflict happen?'. Answering this initial question introduced the potential issues involved in collecting quantitative data that is less available and accessible. As well as data, the literature on community conflicts in Indonesia is often more accessible within the country.

Research relevant to social conflicts, particularly in the context of Indonesia, increasingly uses a case study approach (Barron *et al.*, 2009). Although case study research can focus on a single case or multiple cases, the literatures suggested that a holistic approach would be best to explain the situation (Hammersley and Gomm, 2000). In order to be able to capture the richness and complexity of the issues in this topic, research could potentially collect and analyse qualitative data (Sutrisna, 2009), which would mostly rely on in-depth interviews.

The literature has demonstrated the advantages of using GTM to support the case study method. GTM has the potential to provide an opportunity to obtain primary data, which relies on the advantages of its 'inductive process' (e.g. Glaser and Strauss, 1967). This research proposed a GTM modified to GTS with a proposition of two key elements: the use of literature review to generate theoretical sensitivity and the use of GTS as a method specifically to analyse the data.

This chapter also presented an overview of how GTS can particularly work with continuous comparison in multiple cases. This research approached the cross-case analysis by developing categorisation based on data across three cases to create theoretical sampling. The categorisation appearing from these data components then expanded exponentially to the point of saturation level (e.g. Charmaz, 2006; Corbin and Strauss, 2008). The use of GTS in support of the case study approach revealed that social conflicts – particularly at the communal level in Solo, Poso and Sambas – present a complex relationship between various elements. For example, the complexity of the relationship between ethnicity and religion could become problematic when they interact with other elements, such as personal-level and city-level elements that are criminal, economic or political. A strong negative stereotype can often develop because one community or sociocultural group has a strong association with negative behaviour, or even criminal activity. When this happens repetitively, one social group will develop particular perceptions of another social group.

In most cases, the relationship between sociocultural groups has influenced and has been influenced by the urban-development pattern. For example, the community becomes divided along ethnic, religious and socio-economic lines. In terms of physical settings, the communities then prefer to be segregated and live only within their community. In some cases, segregation was unavoidable, as this has been happening for a long time before the country gained its independence. Nonetheless, the common response to the current conflicts has led to further segregation. Although segregation often results in two or more separate sociocultural groups, the findings recognise the danger of socio-economic segregation in the future. The findings exemplify this with the further links identified to issues of development distribution between rural and urban areas, antisocial behaviour and economic development.

Thus, social diversity has become a potential source of communal conflicts. On some levels, these dynamics have a reciprocal relationship with urban development. Urban development has helped to shape and exaggerate differences within the society, but then the communal conflicts have led to problems in urban development. For example, conflicts have forced a number of communities to leave their homes. The urban centre lost its inhabitants, and some buildings became vacant. These community members also had to seek and build new homes, which were sometimes located in natural-reservation areas.

This complex relationship emerged from interviews with forty-two participants from four main stakeholder groups, as shown by the two examples in Figures 15.4 and 15.5. The emerging issues formed twenty-nine subcategories, which were later condensed into eight main categories: the story behind conflicts, implication and precedent, personal conflicts, cultural elements, community development, economic development, urban planning and development, and governance (Figure 15.6). These eight categories resulted in three layers of development: personal, community and urban.

References

Abdulrahman, H., Darwis, Mahid, S. and Haliadi (2003). *Sejarah Poso*. Yogyakarta, Indonesia: Tiara Wacana.

Arafat, Y. (1998). *Konflik dan Dinamika Etnik Dayak-Madura di Kalimantan Barat: Kajian Melalui Perspektif Komunikasi Antaretnis, Prasangka Sosial, dan Kepentingan Politik pada Kerusuhan Antara Etnis Daya dengan Etnis Madura di Kalimantan Barat*. Master (S2) thesis, Progrom Studi Sosiologi – Jurusan Ilmu-ilmu Sosial. Gadjah Mada University, Yogyakarta, Indonesia.

Barrett, P. S. and Barrett, L. C. (2003). 'Research as a kaleidoscope on practice', *Construction Management & Economics*, 21: 755–66.

Barron, P., Kaiser, K. and Pradhan, M. (2009). 'Understanding variations in local conflict: Evidence and implication from Indonesia', *World Development*, 37: 698–713.

Bertrand, J. (2008). 'Ethnic conflict in Indonesia: National models, critical junctures, and the timing of violence', *Journal of East Asian Studies*, 8: 425–49.

Bollens, S. A. (2006). 'Urban planning and peace building', *Progress in Planning*, 66: 67–139.

Cahyono, H., Tryatmoko, M. W., Adam, A. W. and Satriani, S. (2008). *Konflik Kalbar dan Kalteng: Jalan Panjang Meretas Perdamaian*. Yogyakarta, Indonesia: P2P-LIPI – Pustaka Pelajar.

Charmaz, K. (2006). *Constructing Grounded Theory: A practical guide through qualitative analysis*. London: Sage.

Colombijn, F. and Lindbald, J. T. (2002). *Roots of Violence in Indonesia*. Leiden, Netherlands: KITVL (Koninklijk Instituut voor Taal-, Land- en Volkenkunde) Press, Royal Institute of Linguistic and Anthropology.

Coppel, C. A. (2006). Violence: Analysis, represetation and resolution. In C. A. Copel (ed.), *Violent Conflict in Indonesia: Analysis, representation, resolution* (pp. 3–18). Abingdon, UK: Routledge.

Corbin, J. and Strauss, A. (2008). *Basics of Qualitative Research: Techniques and procedures for developing grounded theory*. London: Sage.

Creswell, J. W. (1994). *Research Design: Qualitative and quantitative approaches*. London: Sage.

Dijk, K. v. (2002). The good, the bad and the ugly, explaining the unexplainable: Amuk Massa in Indonesia. In F. Colombijn and J. T. Lindbald (eds), *Roots of Violence in Indonesia* (pp. 278–97). Leiden, Netherlands: KITVL (Koninklijk Instituut voor Taal-, Land- en Volkenkunde) Press, Royal Institute of Linguistic and Anthropology.

Drexler, E. (2006). Provoking violence, authenticating separatism: Aceh's humanitarian pause. In C. A. Chopel (ed.), *Violent Conflict in Indonesia: Analysis, representation, resolution* (pp. 163–73). New York: Routledge.

Glaser, B. G. and Strauss, A. L. (1967). *The Discovery of Grounded Theory: Strategies for qualitative research*. New Brunswick, NJ: Aldine Transaction.

Goulding, C. (2002). *Grounded Theory: A practical guide for management, business and market researchers*. London: Sage.

Grix, J. (2010). *The Foundation of Research*. Basingstoke, UK: Palgrave Macmillan.

Hammersley, M. and Gomm, R. (2000). Introduction. In M. Hammersley, R. Gomm and P. Foster (eds), *Case Study Method* (pp. 234–58). London: Sage.

Hancock, D. R. and Algozzine, B. (2006). *Doing Case Study Research: A practical guide for beginning researchers*. New York: Teachers College Press.

Heath, H. and Cowley, S. (2004). 'Developing a grounded theory approach: A comparison of Glaser and Strauss', *International Journal of Nursing Studies*, 41: 141–50.

Kataoka, M. (2010). 'Factor decomposition of interregional income inequality before and after Indonesia's economic crisis', *Studies in Regional Science*, 40: 1061–72.

Koeswinarno and Abdurahman, D. (eds) (2006). *Fenomena Konflik Sosial di Indonesia: Dari Aceh sampai Papua*. Yogyakarta, Indonesia: Lembaga Penelitian Universitas Islam Negeri Sunan Kalijaga.

Koeswinarno, Setyo, B., Dwiyanto, A. and Abdullah, I. (2004). Resolusi Konflik dalam Perspektif Sosial Budaya. Hibah Pekerti Angkatan I 2003-2004. Direktorat Penelitian dan Pengabdian pada Masyarakat. Direktorat Jenderal Pendidikan Tinggi – Departemen Pendidikan Nasional, Jakarta.

Kurian, G. T. (2001). *The Illustrated Book of World Ranking* (5th edn). New York: M. E. Sharpe.

Loveband, A. and Young, K. (2006). Migration, provocateurs and communal conflict: The cases of Ambon and West Kalimantan. In C. A. Copel (ed.), *Violent Conflict in Indonesia: Analysis, representation, resolution* (pp. 144–62). Abingdon, UK: Routledge.

McCreaddie, M. and Payne, S. (2010). 'Evolving grounded theory methodology: Towards a discursive approach', *International Journal of Nursing Studies*, 47: 781–93.

McLaughlin, K., and Ari, P. (2010). *Conflict and Dispute Resolution in Indonesia: Indonesian Social Development Paper 16*. Jakarta: World Bank.

Mancini, L. (2008). Horizontal inequality and communal violence: Evidence from Indonesia districts. In F. Stewart (ed.), *Horizontal Inequalities and Conflict: Understanding group violence in multiethnic societies* (pp. 106–35). Basingstoke, UK: Palgrave Macmillan.

Muller, C. and Vothknecht, M. (2011). Group Violence, Ethnic Diversity and Citizen Participation: Evidence from Indonesia. Marseille, France: Centre de Recherche en Developpement Economique et Finance Internationale (DEFI), Universite de la Mediterranee Aix-Marseille II.

Murtagh, B., Graham, B. and Shirlow, P. (2008). 'Authenticity and stakeholder planning in the segregated city', *Progress in Planning*, 69(2): 41–92.

Onwuegbuzie, A. J. and Collins, K. M. T. (2007). 'A typology of mixed methods sampling designs in social science research', *The Qualitative Report*, 12: 281–316.

Panggabean, R. (2006). Pola Kekerasan Kolektif di Indonesia 1990–2003. In: Koeswinarno and D. Abdurahman (eds), *Fenomena Konflik Sosial di Indonesia: dari Aceh sampai Papua* (pp. 83–138). Yogyakarta, Indonesia: Lembaga Penelitian Universitas Islam Negeri Sunan Kalijaga.

Peluso, N. L. (2008). 'A political ecology of violence and territory in West Kalimantan', *Asia Pacific Viewpoint*, 49: 48–67.

Poerwanto, H. (2006). Konflik dan Masalah Identifikasi Diri Orang Cina. In Koeswinarno and D. Abdurahman (eds), *Fenomena Konflik Sosial di Indonesia: dari Aceh sampai Papua* (pp. 216–41). Yogyakarta, Indonesia: Lembaga Penelitian Universitas Islam Negeri Sunan Kalijaga.

Purdey, J. (2006). The 'other' May riots: Anti-Chinese violence in Solo, May 1998. In C. A. Coppel (ed.), *Violent Conflict in Indonesia: Analysis, representation, resolution* (pp. 72–89). Abingdon, UK: Routledge.

Saunders, M., Lewis, P. and Thornhill, A. (2009). *Research Methods for Business Students* (5th edn). Harlow, UK: Pearson Education.

Strauss, A. L. (1987). *Qualitative Analysis for Social Scientist* (14th printing 2003). Cambridge, UK: Cambridge University Press.

Strauss, A. L. and Corbin, J. (1998). *Basics of Qualitative Research: Techniques and procedures for developing grounded theory* (2nd edn). Thousand Oaks, CA: Sage.

Sutrisna, M. (2009). Research Methodology in Doctoral Research: Understanding the meaning of conducting qualitative research. *ARCOM Doctoral Workshop*. Liverpool, UK.

Sutrisna, M. and Barrett, P. (2007). 'Applying rich picture diagrams to model case studies of construction projects', *Engineering, Construction & Architectural Management*, 14: 164–79.

Tadjoeddin, M. Z. and Murshed, S. M. (2007). 'Socio-economic determinants of everyday violence in Indonesia: An empirical investigation of Javanese districts, 1994–2003', *Journal of Peace Research*, 44: 689–709.

Vaus, D. d. (2001). *Research Design in Social Science*. London: Sage.

Yin, R. K. (2009). *Case Study Research: Design and methods* (4th edn). London: Sage.

Part VII

Design science research

16 Design science methodology for developing a learning object repository for construction

Pathmeswaran Raju and Vian Ahmed

The arrival of new technologies makes learning content out of date and thus often leads to reinvention of the wheel. The concept of learning objects is the basis for a new instructional design paradigm for e-learning that puts the emphasis on reuse as a key characteristic of learning activities. In the last decade, several efforts towards the standardisation of learning technology have resulted in the emergence of specific terminologies that are used to name and classify learning elements. Most of these efforts underline the importance of providing metadata information for content in a standardised format as a crucial step to reusability. Most of the current learning object repositories provide room for metadata, but lack a semantic model that clearly establishes what a learning object is and what its associated metadata such as pedagogy should be. The lack of semantics in learning objects also affects their interoperability, sharability and reusability. The Semantic Web-based solution facilitates the easy integration and identification of meaningful interrelationships, represents and uniquely identifies the digital resources and improves search precision. However, Semantic Web has not been fully applied to e-learning, particularly for developing and delivering reusable learning objects. Therefore, this research brings novelty by developing an ontology for the construction domain in order to share and reuse construction learning objects. Furthermore, it follows design science research methodology and develops an ontology-driven, Semantic Web-based learning object repository for construction by integrating pedagogical concepts within the construction education ontology. Results from the evaluation indicated that the acceptance rate of the repository is low in the academic setting. Therefore, strategies for the acceptance of innovation by potential users and communities have to be developed in order to improve the acceptability of the learning object repository.

16.1 Introduction

Learning is a constructive process of actions, including solving problems, engaging in dialogues of enquiry and acquiring new knowledge within an environment and reflecting upon it (Sharples, 2000). E-learning (electronic learning) is a general term referring to learning enhanced by computers. It is networked, which makes it capable of instant update, storage, retrieval and sharing of instructions or information. However, rapid development of technologies provokes the need for reinvention of the wheel, as existing e-learning applications become out of date. In the last decade, learning objects, which are smaller chunks of learning content, have received a lot of interest as the basis of a new type of computer-based instruction, in which the instructional content is created from individual components. The concept of learning object has evolved from the need to reuse digital learning materials. Learning objects offer economic as well as pedagogical advantages over traditional learning materials. The learning objects are created just once, but are used several times in different contexts, compensating for the high cost of production. This object-based principle is based upon the idea that a course or lesson can be built from reusable instructional components that can be built separately but modified to the user's needs.

A learning object is a self-contained component with associated metadata that allow the object to be reused in different contexts. Additionally, learning objects are generally understood to be digital entities, deliverable over the Internet, making them accessible and usable by multiple users in parallel (Wiley, 2001). In order for widespread reuse to be possible, interoperability issues are extremely important, and a somewhat neglected kind of interoperability is interoperability between learning objects (Duval and Hodgins, 2003). Examples include content objects from different original creations working together when assembled into a learning object and learning objects being able to work properly when moved among systems using different specifications. The learning objects that can be reused within various learning environments have been a core feature of e-learning systems. However, the concept of learning objects has been criticised for being too inflexible and not taking into account the particular learner's learning needs, goals and preferences (Matthews, 2005).

According to Downes (2004), learning objects have five characteristics, namely they are sharable, digital, interoperable, modular and discoverable, and their discoverability is made possible by the learning object metadata (LOM). Learning objects are stored in learning object repositories, of which there are two kinds: one contains learning object and metadata, and the other contains only metadata. Although the definition of the IEEE LOM standard has led to a wide adoption of LOM, learning objects still suffer from how difficult it is to create metadata. Although reusing existing multimedia resources and avoiding duplication save time and money, the effort required to identify, catalogue, store, search, retrieve and finally reuse a learning object is still significant (Duval, 2004; Downes, 2005; Oliver, 2005). There are also several problems that obstruct the effective and efficient reuse of learning objects in education and training (Duval and Hodgins, 2003); among the main problems are the following: (1) enabling large-scale reuse requires the availability of a vast number of learning objects, and (2) finding relevant learning objects in learning object repositories is not always straightforward. The research proposed a conceptual framework for developing an online environment of learning objects, as shown in Figure 16.1. The framework addresses three main challenges in order to produce reusable learning objects, such as the need to make them: (1) intelligent by developing semantic metadata, (2) sharable through content packaging and (3) dynamic using ontologies and Semantic Web.

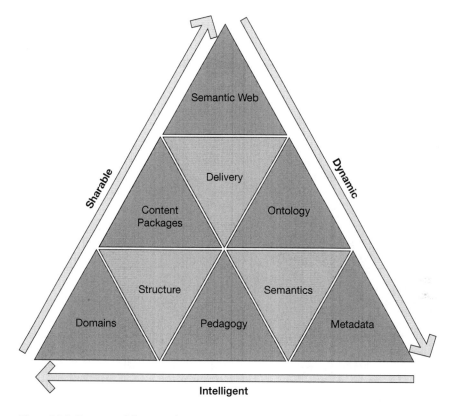

Figure 16.1 Conceptual framework

In order to develop intelligent learning objects, metadata, pedagogy and domains have to be considered as enablers. Similarly, domains, content packaging and Semantic Web make learning objects more sharable. Semantic Web, ontology and metadata have to be utilised for dynamic learning objects to be developed. The discoverability of the learning objects can be enhanced by their being described using a common metadata standard and it being stored in a dynamic, semantic learning object repository that has semantic search capabilities. The reusability of the learning objects can be increased by the production of learning objects that are intelligent and can be packaged together to work in other learning management systems. When a virtual learning object repository is huge and is distributed on the Internet, the use of metadata and keyword searches is inefficient and ineffective, as many potential associations with various learning aspects are bypassed (Mustaro and Silveira, 2006). This has led to Semantic Web approaches that model the relationships between learning objects using formal ontologies (Sicilia and Lytras, 2005). The chapter is set out to discuss the research methodology, data collection and analysis, and the findings from the research conducted to develop the learning object repository using the above conceptual framework (Raju, 2009).

16.2 Research methodology

Research philosophy

Research philosophy relates to the development of knowledge and the nature of that knowledge (Saunders *et al.*, 2007). It contains important assumptions about the way in which the researcher views the world, particularly the assumptions concerning the researcher's view on the relationship between knowledge and the process by which it is developed. The research undertaken for the study aims to develop an artefact, the learning object repository. Thus, this is developmental and technological research that requires an appropriate research methodology. The design science methodology acknowledges IT research as a component of improving and developing artefacts for the development of better solutions (Vaishnavi and Kuechler, 2007). Design science research, by definition, changes the state of the world through the introduction of novel artefacts.

Epistemology is the study that explores the nature of knowledge. It is about what constitutes acceptable knowledge in a particular field of study (Saunders *et al.*, 2007). Ontology is the study that describes the nature of reality. This relates to the assumptions that researchers have about the way the world operates. Axiology is about the study of values: it is concerned with judgements about the researcher's values to see whether the researcher's own values play a part in the stages of the research process (Saunders *et al.*, 2007). Epistemologically, the design science researcher knows that a piece of information is factual and knows further what that information means through the process of construction and circumscription. The learning object repository is constructed, and its behaviour is the result of interactions between components. Descriptions of the interactions are information, and, to the degree the repository behaves predictably, the information is true. Axiologically, the design science researcher values creative manipulation and control of the learning object repository in addition to more traditional research values such as the pursuit of truth or understanding of the pedagogy behind the repository.

Research strategy

Research strategy is the research approach taken towards data collection and analysis. The choice of research strategy is guided by the research objectives, the extent of existing knowledge, the amount of time and other resources available and the philosophical underpinnings (Saunders *et al.*, 2007). This research has chosen design science as its main research strategy because the aim of the research is to develop an artefact for providing better solutions. Design science research is a science of the artificial. It is a body of knowledge about artificial objects and phenomena designed to meet certain desired goals. The intention of natural science is to obtain understanding, whereas that of design science is concerned with achieving human goals by applying this understanding (Haynes, 2001). Simon (1996) describes the natural sciences as those concerned with how things are and how things work; engineering sciences, on the other hand, are concerned with artefacts, the properties of effective artefacts, and how to design them. An essential element of the design sciences is their emphasis on the normative and prescriptive, on improving artefacts through the development of better 'should' and 'ought' statements (Simon, 1996).

16.3 Data-collection strategy

The selection of design science is supported by Vaishnavi and Kuechler (2007), as the reason that design science research is applicable to IT research is because some types of research question occur naturally in the field. This study also adopted a quantitative research strategy during the pre-development and post-development stages. Surveys in the form of questionnaires are used during the requirements capture to identify the user needs and requirements of the learning object repository. During the user trials of the learning object repository, questionnaires and interviews are carried out to evaluate the prototype. The research approach follows the general design cycle, as shown in Figure 16.2 and described by Vaishnavi and Kuechler (2004). In this model, all design begins with awareness of a problem. Design science research is sometimes called "improvement research," and this designation emphasises the problem-solving or performance-improving nature of the activity (Vaishnavi and Kuechler, 2007). Suggestions for a problem solution are abductively drawn from the existing knowledge or theory base for the problem area (Pierce, 1935, cited in Vaishnavi and Kuechler, 2007).

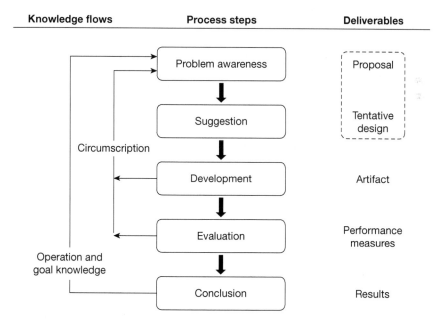

Figure 16.2 General design cycle

Source: Takeda *et al.*, 1990

Vaishnavi and Kuechler (2007) describe the design cycle in detail. An attempt at implementing an artefact according to the suggested solution is development. Partially or fully successful implementations are then evaluated according to the functional specification implicit or explicit in the suggestion. The development, evaluation and further suggestions are often iteratively performed in the course of the design research. The basis of the iteration, the flow from partial completion of the cycle back to awareness of problem, is indicated by the circumscription arrow. Conclusion indicates termination of the project. This research study follows the design cycle in order to develop the learning object repository.

Data-collection methods

The process steps in the design science cycle have outputs at each step, as shown in Figure 16.2. In the first step of the process, a proposal will be an output of the problem awareness. For this research, the research proposal addresses the problem of sharing and reusing learning objects within learning environments. The suggestion is to develop content in terms of learning objects that can be reused and shared among different platforms and applications. A conceptual framework is produced as an output of the suggested solution. The framework addresses three challenges of reusing learning objects, such as dynamic, intelligent and sharable learning objects. The first two steps of the design science life cycle – problem awareness and suggestion – can be considered as the pre-development stage, and the last two steps of the process – evaluation and conclusion – can be considered as the post-development stage. The development stage can be divided into system analysis and design and system development.

Pre-development

The first step of the design science cycle is gaining awareness of a problem through problem identification and definition. The problem identified in the current research is the difficulty in reusing and sharing learning content among educators, learners and curriculum developers. The research attempts to develop a framework that addresses the challenge of developing learning objects that can be reused and shared among various learning platforms. A literature review was carried out to facilitate comprehension of existing theories and work by others, to form a coherent argument for further research, and to demonstrate a fundamental understanding of e-learning. Once the problem has been identified, research is necessary to derive suggestions to address the research problem. Metadata standards provide the labels for the learning objects that enhance the discovery of learning objects within the learning object repository. Content packages provide the ability to package the relevant learning objects together in order to export to a learning management system, such as Blackboard. Learning styles, learning object types and construction domains have been identified so that the three challenges can be addressed. Riding's learning style has been chosen as a suitable style for the development of learning objects. UKLOM and SCORM standards have been chosen as appropriate for metadata tagging and content packaging, respectively. Web standard, in particular Semantic Web, has been explored, with its core technologies, RDF, OWL and SPARQL.

Development

The literature review of learning objects, educational standards and Semantic technologies provides suggestions to address the research problem of the reusability and sharability of learning objects. With the knowledge gained in the first two steps, the next step is to utilise it to implement the suggestion, as discussed in the suggestion phase. This development phase is where most of the actual design takes place, which is the creative effort required to synthesise existing knowledge and a well-defined problem definition into an artefact to solve the problem. A resulting artefact of design science research may be rather abstract in nature, such as in the form of constructs, models or methods (March and Smith, 1995). However, the research has developed a fully working prototype with the functionalities that are required within the repository. Construction education ontology has been developed using Protégé. The programme–module–topic structure has been followed to provide a structure for the learning objects. The learning object types, such as demonstration, have been modelled by consideration of their associated learning style, imager within the ontology. The learning object repository has been developed on the Java platform using NetBeans. The MySQL database has been used to store the metadata of the learning objects. The Jena Semantic Web toolkit has been used to develop the backend of the learning object repository, and a Lucene search engine has been implemented within the learning object repository to provide search mechanisms for learning objects.

Post-development

Evaluation is concerned with gathering data about the usability of a design or product by a specific group of users for a particular activity within a specified environment or work context (Preece *et al.*, 1994). The evaluation of designed artefacts typically uses methodologies available in the knowledge base. A design artefact is complete and effective when it satisfies the requirements and constraints of the problem it was meant to solve (Hevner *et al.*, 2004). The choice of design evaluation method can vary, depending on the designed artefact and the selected evaluation metrics. Experimentation – in particular, control experiments – has been chosen as an evaluation method for the research, as it has high internal validity and control (Whitley, 1996). As a first method of evaluation, a controlled experiment is carried out to test the usability of the system. Experiments in terms of testing the functionality, usability and user acceptability are used to evaluate the learning object repository. Functional (black-box) and structural (white-box) tests are carried out to check the systems for any failures in execution of any commands during the software testing. The evaluation examines whether the learning object repository meets the pedagogical needs of educators and learners. The evaluation also tests the functionality and usability of the repository using a standard usability questionnaire.

Usability tests are carried out in order to test the system for its fitness for purpose (effective, efficient and satisfying) in the context of use (user, tasks, sociotechnical environment; McClelland, 1995). In order for a usability test to be carried out, a small sample of users is selected from the user group. The learning object repository has been tested and evaluated at this stage. In order to identify the defects and failures of the repository, functional (black-box) and structural (white-box) tests have been carried out. Evaluation in terms of user trials has been conducted to measure the effectiveness, efficiency and satisfaction of the repository. The fourteen users (five educators, seven learners and two learning technologists) participated in the user trials. A system usability scale (SUS) has been used to test the usability of the learning object repository. Participants were asked to perform three tasks within the repository, namely submit a learning object, search for a learning object and create a content package. Data were analysed by three methods: total score, maximum rating and goal achievement. Interviews were also conducted to identify whether the repository meets the pedagogical needs of the users and to measure the acceptability of system in the academic environment.

16.4 Data analysis

After the development of an artefact, it is necessary to evaluate the artefact using empirical methods 'to determine how well an artefact works' (Hevner *et al.*, 2004). According to Hevner *et al.* (2004), IT artefacts can be evaluated in terms of functionality, completeness, consistency, accuracy, performance, reliability, usability, fit with the organisation and other relevant quality attributes. Thus, the learning object repository has been evaluated for its functionality, completeness, performance, usability and user acceptance. The functionality, completeness, performance and usability of the system are measured by the questionnaire. The user acceptance of the system is examined by the user interviews. Moreover, Handy *et al.* (2001) developed a model for user acceptance that demonstrates how the individual, system and organisational characteristics affect the acceptability of the system. Individual characteristics relating to the learning object repository have been tested by interviews. System characteristics such as 'perceived usefulness', 'perceived ease of use' and 'perceived system acceptability' are tested by questionnaire.

User trial requires a sample of users that in some way reflects the product user population as whole. According to McClelland (1995), this means selecting a group of users who don't just have the same characteristics as the user population, but who also reflect the extent to which these characteristics vary. However, it is important that the characteristics of the users are not just the physical or psychological aspects, but are also related to the aspects of the interaction to be examined in the trial. There are two aspects to be considered when the sample population is chosen. The first is to develop the profile of the users in the population, and the second is to get the appropriate sample numbers. For the purpose of the user trial of the learning objects repository, the main profiles of the

Table 16.1 Population sample of user trials

Profile of users	Numbers of users
Educators	5
Learners	7
Curriculum developers	2
Total	14

users identified are educators, learners and curriculum developers. The decision on the appropriate number of users for the user trial is always a difficult issue. Virzi (1992, cited in McClelland, 1995) concluded that 80 per cent of the usability problems can be detected by four or five subjects, as more subjects tend to detect fewer and fewer insights as numbers increase. However, Tullis and Stetson (2004) tested the usability of a system with three kinds of scale: SUS, questionnaire for user interface satisfaction (QUIS) and computer system usability questionnaire (CSUQ). They found that most of the questionnaires appear to reach an asymptote at a sample size of twelve. However, fourteen users participated in this trial owing to three different sets of user profiles. Table 16.1 shows the breakdown of the users, who were from the School of the Built Environment at the University of Salford. Prior to each user trial, participants were given a full background to the research and had the research methodology explained. Participants were also informed that they were free to withdraw from the trial at any time, without explanation or prejudice.

Phase 1: Questionnaire

The questionnaire technique is used to examine the usability of the learning object repository. Brooke's (1996) SUS questionnaire is used for the evaluation, as it is relatively short, but has been proved to be more effective than longer counterparts such as QUIS and CSUQ (Tullis and Stetson, 2004). The SUS is a simple, ten-item attitude scale giving a global view of subjective assessments of usability. According to ISO standard ISO 9241 Part 11, the usability of a system can be measured only by taking into account the context of use of the system (ISO, 1998) – that is, who is using the system, what they are using it for and the environment in which they are using it.

The SUS scale is used after the respondent has had an opportunity to use the system, but before any discussion takes place. Respondents are asked to record their immediate response to each item, rather than thinking about items for a long time. Respondents are also asked to check all the items, and, if a respondent feels that they cannot respond to a particular item, they should mark the centre point of the scale. The evaluator briefs the respondents with the purpose of the learning object repository, and then a set of tasks are given to them to perform with the system. The evaluator also employs the procedure of 'think aloud', followed by an interview. Think aloud is a process used during evaluations in which the users are encouraged to talk to the evaluator and discuss freely the problems they are having with the system (Baillie and Schatz, 2006). By using this technique, the users see themselves as collaborators in the evaluation and not simply as experimental subjects. This also allows the evaluator to ask the users questions. According to Dix *et al.* (1998), this form of evaluation has two advantages: the user is encouraged to criticise the system, and the evaluator can clarify points of confusion at the time they occur and so maximise the effectiveness of the approach for identifying problem areas. The set of tasks that was given to respondents is given below.

- Task 1: Submit a learning object.
- Task 2: Search for a learning object, which consists of simple search, advanced search and browse search.
- Task 3: Create a content package.

The usability test is set to measure three aspects – effectiveness, efficiency and satisfaction:

- effectiveness: a task is completed successfully;
- efficiency: the time taken to achieve a goal;
- satisfaction: the results of questionnaires.

In order to measure efficiency, the time taken to achieve a goal is measured from the time the user is asked to start, to the time when the goal has been achieved. The expected time to complete the task has been estimated from pre-tests. According to Bevan (2007), the maximum time allowed to users before they are categorised as having failed should be at least three times the expected time. Table 16.2 outlines the allocated time for each task, with the expected time.

Method 1: Total score

The usability testing using SUS is carried out with fourteen users to test the satisfaction of the users in using the learning object repository. The SUS score is calculated by summing up the score contributions from each item. Each item's score contribution ranges from 0 to 4. For Items 1, 3, 5, 7 and 9 in the SUS, the score contribution is the scale position minus 1. For Items 2, 4, 6, 8 and 10, the contribution is 5 minus the scale position. The total SUS score is obtained by

Table 16.2 Time allocated for tasks

Tasks	Expected time (minutes)	Allocated time (minutes)
Submit a learning object	3	9
Search for a learning object	4	12
Create a content package	3	9

multiplying the sum of the scores by 2.5 to obtain the overall value of SUS. The SUS score has a range of 0–100, which represents a composite measure of the overall usability of the system being studied, in this case, the learning object repository. Figure 16.3 shows the results of the usability tests with all fourteen users. All the questionnaires score above 85 per cent, with one test scoring 100 per cent. The figures show high satisfaction with using the learning object repository among the respondents.

Method 2: Maximum rating

Questionnaires were converted to percentages by dividing each score by the maximum score possible on that scale. The frequency distributions of the ratings on each questionnaire are converted to percentages, as described above, and are shown in Figure 16.4. The higher frequencies of the maximum rating in the SUS responses indicate that users are satisfied with the learning object repository, its interface and functionalities.

Method 3: Goal achievement

SUS questionnaire respondents are allocated a maximum amount of time for successful goal achievement for each task, as set out in the criteria in Table 16.2. Figure 16.5 shows the goal achievement of the respondents. All the users completed

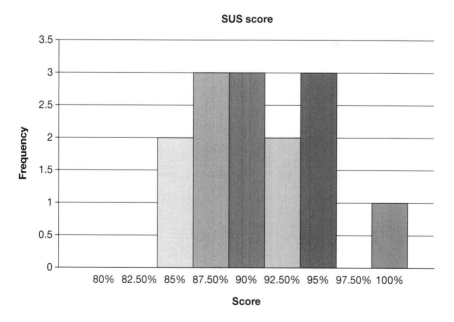

Figure 16.3 Usability test SUS scores

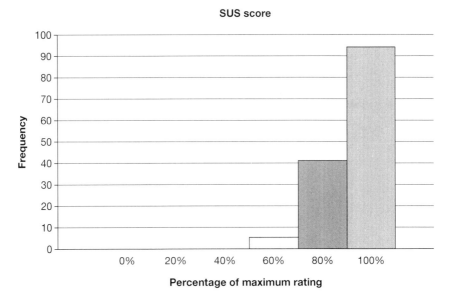

Figure 16.4 Maximum rating in SUS questionnaire

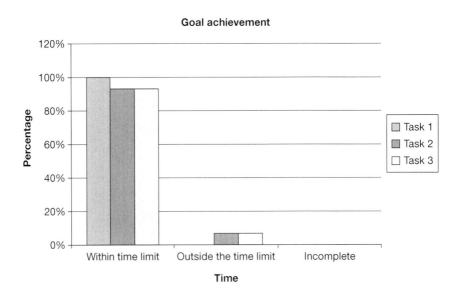

Figure 16.5 Goal achievement by time

Task 1 successfully within the allocated time; 93 per cent of the users completed Tasks 2 and 3 within the allocated time; and only 7 per cent of the users completed Tasks 2 and 3 outside the allocated time. It shows high efficiency of the learning object repository in terms of its usage and functionalities. In addition to that, all the users completed the tasks in the usability tests, which demonstrates 100 per cent effectiveness of the learning object repository.

Phase 2: Interviews

Most user trials involve some form of interview as a way of asking users their general opinion of the user trial and system. However, the interview should have clear structure and direction, and the evaluator should be prepared to prompt and question the user in depth, if necessary (McClelland, 1995). In the trial of the learning object repository, the same users who participated in the questionnaire method of user trial were interviewed with two sets of questions. The first set was to evaluate the learning objects repository for pedagogical needs. The questions were aimed at developing understanding of whether the learning objects repository meets the pedagogical needs of educators and learners. The second set of questions was aimed at developing understanding of the acceptability of the repository in the academic environment. Table 16.3 summarises the responses received from the users to the questions on pedagogical needs. The responses indicate that users

Table 16.3 Summary of the responses on pedagogical needs

Questions	Summary of answers
What motivates you to use the system?	Ease of finding relevant learning objects; use of ontology to structure programmes, modules and topics; content packages to export the learning objects to Blackboard; ease of finding learning objects on the topics I want
Do you perceive the benefits of the system?	Yes, students can benefit most; depends on the availability of the learning objects; provides some benefits, but takes time to develop learning objects; could see the benefits, but need support; it has some benefits, but there are too many e-learning tools: I need some training
Are the pedagogical ontologies meaningful and useful?	Yes, useful to find suitable learning objects; I would expect learning outcomes as an element of the pedagogy; yes, but some topics are not available in the ontology; it is useful, but can I add new topics?; I think it is useful, but not sure about meaningful: some topics, lessons are confusing
How do you feel about sharing your materials and using others' materials?	I will share and may use others' materials; I would share mine, but not sure about others' objects; copyright issue can be a problem; no validity/creditably of others' content; someone has to convert my lecture notes to learning objects
How could you make use of this environment?	As a parallel to Blackboard; for depositing PowerPoint/lecture notes; to find some definitions/diagrams on the topic I need; can be used to prepare course materials; I can find some literatures on a topic I want; I can put up some of my old lecture notes

Table 16.4 Summary of the responses on acceptability

Questions	Summary of answers
Do you find this system eases your workload?	I don't think so, still need to develop course materials (educator); not sure (learner, educator); maybe it will ease, I can easily find the relevant materials (learner); it could if everybody uses it (curriculum developer); yes, if students are willing to get lectures notes/material from the repository (educator); not really, I would need to develop objects for the repository if it was ever used in our school (curriculum developer)
Do you think use of repository will improve teaching and learning?	Teaching and learning will improve; however, it has to be integrated within school strategy; yes, it will definitely improve teaching and learning, but who is going to invest time in this?; depends on how you use it, clear mechanisms need to be developed in order to use it effectively; will help to improve teaching, but how are you going to find the objects?; I don't think any lecturers will use this
Are you willing to use this repository?	I will, if necessary (learner); I will use it, but not as a core teaching tool (educator); will use, but not often (learner); possibly, can be used for creating content packages in our Blackboard (curriculum developer); yes, if I get time (educator); not willing, it will need some additional time

understand the system, its purpose and its benefits. However, copyright issues are raised about the question of sharing the learning objects within the environment.

The responses indicate that the learning object repository meets the pedagogical needs of the educators, learners and even curriculum developers; however, it is necessary to understand the acceptability of the system within the educational environment. As part of the user interviews, three questions were posed to find the acceptability of the system within the educational context. Table 16.4 provides a summary of the responses received from the interviews.

Responses indicate that, although users understand the benefits of the system, they don't find that the system will ease their workload. Some users believe that it will take some time to get started with the system in the learning environment. Although users are willing to use the repository, there are concerns about submission of learning objects to the repository and their limited time allocation for e-learning tools.

16.5 Main findings and conclusions

Although many e-learning tools are developed for construction education, many of these tools often become redundant and technically out of date, raising the need to reinvent the wheel. Producing standardised learning objects with pedagogical concepts attached to them is the way forward to avoid reinventing the wheel. The research proposed a conceptual framework for developing dynamic, intelligent and sharable learning objects. The framework brought enabling standards and technologies together to develop a dynamic learning object repository. The framework is generic and can be utilised to develop a repository of sharable learning objects for any field. The framework was tested by a prototype learning object repository for construction being developed.

A metadata standard offered semantic annotation for learning objects and thus enhanced the discoverability and reusability of learning objects. UKLOM was adapted to develop the learning object repository as it provided the context of UK education. There are no existing metadata for construction learning objects, and therefore this research proposed a metadata framework for construction by integrating the construction domain as an element of the UKLOM standard. The application of e-learning standards facilitated the development of interoperable learning objects and the dynamic learning object repository that provided access to large sets of learning objects. The discoverability of the learning objects was enhanced by the use of metadata to make the learning objects more intelligent and also by the development of the learning object repository on the Semantic Web. The reusability of learning objects has been achieved by developing dynamic, intelligent learning objects using metadata and content packaging standards and by the use of ontologies and Semantic Web for the development of a dynamic learning object repository.

The design science methodology has been followed to develop the learning object repository, using ontologies on the Semantic Web. Semantic Web is emerging as a next-generation web and has huge potential for developing intelligent learning objects and supporting e-learning at large. Semantic Web and ontologies offer great educational value to curriculum developers and users who are desperate for change in the way traditional e-learning tools and applications work. E-learning pedagogies such as learning styles have been employed effectively within the learning object repository, which can be tailored to suit the needs of the particular learner. The developed learning object repository has been evaluated by users for usability, functionality and acceptability, and to see if it satisfies the pedagogical needs of the users. The repository scored highly in usability and functionality testing; however, the acceptability of such a system is low in the academic setting, for several reasons. Academic institutions should encourage the use of various e-learning tools, rather than concentrating on a particular learning management system (e.g. Blackboard) and should also recognise the successful implementation of such tools in academic settings.

The learning object repository was evaluated for efficiency, performance, functionality and usability; however, it was not evaluated for its effectiveness as a learning tool in the classroom or learning environment. The purpose of the learning objects in the learning object repository is to demonstrate the concept, and, therefore, more learning objects need to be added to the repository in order for it to be used in real-life settings. The selected evaluation sample was limited to one institution, mainly to evaluate the usability and efficiency of the learning object repository. To develop a complete set of vocabulary for construction education, further study is needed to identify further concepts and domains for construction education. This will enable the adaptation of learning objects on a large scale.

In order to measure the effectiveness of the repository, further work is needed to assess learners' skills before and after using the repository. Moreover, further study is needed in order to identify the needs of the industry and professional bodies for developing continuous professional development programmes.

References

Baillie, L. and Schatz, R. (2006). 'A lightweight, user-controlled system for the home', *An Interdisciplinary Journal on Humans in ICT Environments*, 2(1): 84–102.

Bevan, N. (2007). Designing a user-based evaluation [online]. http://citeseerx.ist.psu.edu/viewdoc/download?doi=10.1.1.475.8581&rep=rep1&type=pdf (accessed 6 December 2015).

Brooke, J. (1996). SUS: A 'quick and dirty' usability scale. In P. W. Jordan, B. Thomas, B. A. Weerdmeester and A. L. McClelland (eds), *Usability Evaluation in Industry*. London: Taylor & Francis.

Dix, A., Finlay, J., Abowd, G. and Beale, R. (1998). *Human–Computer Interaction* (2nd edn). Hemel Hempstead, UK: Prentice Hall.

Downes, S. (2004). The Learning Marketplace: Meaning, metadata and content syndication in the learning object economy [online]. Available at www.downes.ca/files/book3.htm (accessed 2 August 2007).

Downes, S. (2005). 'E-learning 2.0', *eLearn Magazine*, 10: 5.

Duval, E. (2004). We're on the road to . . . In L. Cantoni and C. McLoughlin (eds), *Proceedings of World Conference on Educational Multimedia, Hypermedia and Telecommunications 2004* (pp. 3–8). Lugano, Switzerland: AACE.

Duval, E. and Hodgins, W. (2003). 'A LOM research agenda', *Proceedings of the Twelfth International Conference on World Wide Web*, p. 19. ACM Press.

Handy, J., Hunter, I. and Whiddett, R. (2001). 'User acceptance of inter-organizational electronic medical records', *Health Informatics Journal*, 7(2): 103–7.

Haynes, S. R. (2001). 'Explanation in information systems: A design rationale approach'. PhD thesis, London School of Economics and Political Science, Department of Information Systems and Department of Social Psychology.

Hevner, A., March, S., Park, J. and Ram, S. (2004). 'Design science in information systems research', *MIS Quarterly*, 28(1): 75–105.

ISO. (1998). ISO 9241-11. Available at www.iso.org/obp/ui/#iso:std:iso:9241:-11:dis:ed-2:v1:en (accessed 6 December 2015).

McClelland, I. (1995). Product assessment and user trials. In H. Wilson and N. Corlett (eds), *Evaluation of Human Work: A practical ergonomics methodology* (2nd edn, pp. 249–84). London: Taylor & Francis.

March, S. and Smith, G. (1995). 'Design and natural science research on information technology', *Decision Support Systems*, 15: 251–66.

Matthews, B. (2005). Semantic Web technologies. JISC Technology and Standards Watch Report TSW0502 [online]. Available at www.jisc.ac.uk/uploaded_documents/jisctsw_05_02bpdf.pdf (accessed 30 January 2008).

Mustaro, P. N. and Silveira, I. F. (2006). 'Learning objects: Adaptive retrieval through learning styles', *Interdisciplinary Journal of Knowledge & Learning Objects*, 2: 35–46. Available at http://ijello.org/Volume2/v2p035-046Mustaro.pdf (accessed 20 April 2007).

Oliver, R. (2005). 'Ten more years of educational technologies in education: How far have we travelled?', *Australian Educational Computing*, 20(1): 18–23.

Pierce, C. S. (1935). *Collected Papers*. C. Harshorne and P. Weiss (eds). Cambridge, MA: Harvard University Press (1931–1935).

Preece, J., Rogers, Y., Sharp, H., Benyon, D., Holland, S. and Carey, T. (1994). *Human–Computer Interaction* (Part VI, p. 595). Harlow, UK: Addison Wesley.

Raju, P. (2009). 'An ontology-driven approach for developing learning object repository for construction using Semantic Web'. Unpublished PhD thesis, University of Salford, UK.

Saunders, M., Lewis P. and Thornhill, A. (2007). *Research Methods for Business Students* (4th edn). Harlow, UK: Pearson Education.

Sharples, M. (2000). 'The design of personal mobile technologies for lifelong learning', *Computers & Education*, 34(3–4): 177–93.

Sicilia, M. A. and Lytras, M. (2005). 'On the representation of change according to different ontologies of learning', *International Journal of Learning & Change*, 1(1): 66–79.

Simon, H. (1996). *The Sciences of the Artificial* (3rd edn). Cambridge, MA: MIT Press.

Takeda, H., Veerkamp, P., Tomiyama, T. and Yoshikawam, H. (1990). 'Modeling design processes', *AI Magazine*, Winter: 37–48.

Tullis, T. S. and Stetson, J. N. (2004). 'A comparison of questionnaires for assessing website usability', *Proceedings, Usability Professionals Association Conference* (UPA 2004, pp. 7–11), Minneapolis, MN.

Vaishnavi, V. and Kuechler, W. (2004). Design Research in Information Systems [online]. Available at http://desrist.org/desrist/content/design-science-research-in-information-systems.pdf (accessed 6 December 2015).

Vaishnavi V. and Kuechler Jr, W. (2007). *Design Science Research Methods and Patterns: Innovating information and communication technology*. Boca Raton, FL: CRC Press.

Virzi, R. A. (1992). 'Refining the test phase of usability evaluation: How many subjects is enough?', *Human Factors*, 34(4): 457–68.

Whitley, B. E. (1996). Behavioral science: Theory, research, and application. In *Principles of Research in Behavioral Science* (pp. 1–30). Mountain View, CA: Mayfield.

Wiley, D. A. (2001). Connecting learning objects to instructional design theory: A definition, a metaphor, and a taxonomy. In D. A. Wiley (ed.), *The Instructional Use of Learning Objects* [online version, Open Publication License]. Available at http://reusability.org/read/chapters/wiley.doc (accessed 5 December 2015).

Index